Praise For Lynne &

"Running the mountainous width of Wales is an impressive undertaking in shoes, let alone without. Well done." – *Sir Ranulph Fiennes*

"A remarkable lady and a remarkable challenge. An exciting combination." – *Robert Plant, musician*

"Having witnessed Lynne walk her talk and admired the most satisfactory results at a time when reconnecting to nature is pertinent, I believe Barefoot and Before to be the outstanding book of the present time."
– *Howard Marks, best-selling author and public speaker*

"Of all the people to wish well as they prepare to run 52 miles barefoot (not that there can be many with enough balls let alone stamina to do so), I can't possibly imagine anyone I'd want to wish well more than the wonderful, stupendous, magnificent, and utterly inspiring, lovely Lynne."
- *Barefoot Doctor, author, practitioner and teacher of Taoism*

"I just put one foot in front of the other and repeat." - *Lynne Allbutt*

BAREFOOT AND BEFORE

ONE WOMAN, ONE COUNTRY, TWO BARE FEET

The Inspirational Autobiography
By Lynne Allbutt,

The First Person To Run Across Wales Barefoot

ISBN-10:1514234505
ISBN-13:978-1514234501

This first edition published June 2015

Follow Lynne on Twitter
@lynneallbutt

www.lynneallbutt.com
www.allbuttshoes.com

Author's Note and Acknowledgments:

Whilst the bare footing sections of Barefoot and Before are reasonably recent recollections, obviously a large proportion of the 'before' memories are considerably further back in the rather chaotic filing cabinet of my mind. I have endeavoured to recall and relate them to the best of my ability however I am aware that the memories and perception of the very same situation may differ to those of other people involved. Just in case my enthusiasm has occasionally over-ridden my memory recall, please be reassured that every episode has been relayed and delivered with good intention.

I am extremely grateful to those who have contributed to this book through being in my life. Those who are named, those who have been alluded to and those who haven't been referred to at all, I am grateful to you all. I am where I am today because of all of you. Thank you.

And of course gratitude and appreciation must be expressed to those who have helped the book to fruition - those who transformed it from a lot of words on a screen in my living room into what you are now reading. Thank you to Sian Phillips who can be found and followed at www.twitter.com/_Sians and Lisa Powell for their editing and proof reading skills, to Hazel Brewer for designing the cover, to Steve Thomas www.beacons-sky.uk for the cover photographs and to Tom Wood for overseeing it all. Thank you also to RWA e-learning collectively for their endless support and patience.

Finally and fittingly, much respect and gratitude to my mother for her courage to allow me to share what has obviously been a difficult journey for her, and to my brother for keeping my feet on the ground as only a brother can.

Lynne Allbutt

For Bob.

Foreword - Written By Tom Wood

Over the last few years I have had the pleasure of bouncing many ideas around with Lynne on a regular basis. We have similar values around education, creative thinking, change management and transformational leadership which means that whilst we are worlds apart professionally, there are always synergies in our goals and our approach to life. Typically, a lot of these ideas don't stick, but knowing how determined Lynne can be, one always knows that eventually she will grab something and run with it. This is why, when Lynne walked into my office in 2013 and said that she was going to run barefoot across Wales, I knew that it was a) bonkers and b) going to happen.

What followed has not only been an inspirational journey for Lynne, but also an amazing opportunity for the team at RWA who project managed the challenge and who were invaluable to the successes achieved. The young team we put forward all shared Lynne's passion and in return they have acquired so many new skills.

We were extremely lucky to be part of Lynne's journey from the conception of the original idea, to the literal finish line on Ynyslas Beach in 2014 where we witnessed someone who had both physically and mentally pushed themselves to the limit. We all learned a lot during this challenge and each of us took away something that will remain with us forever.

This autobiography, whilst harrowing at times, highlights that this demanding challenge was not just about running across Wales barefoot, it was about overcoming the challenges of life and achieving a goal that may seem terribly out of reach. There are not many who could achieve what Lynne accomplished, but in reading this book you will hopefully begin to understand why it was so important to her that she did.

T.W. 2015

"When the Japanese mend broken objects, they aggrandize the damage by filling the cracks with gold. They believe that when something has suffered damage and has a history, it becomes more beautiful."

- Billie Mobayed

BAREFOOT AND BEFORE

1.
Dipping my toe in the water

'I wanted to be the first to achieve something;
there aren't many firsts left.'

I cursed my trainers. They nipped at my Achilles tendon. My ankles felt unreliable, my knees ached. I was trying to run my way out of a dark mood. Nothing was wrong and yet nothing was right.

I was engrossed in that all too familiar feeling you get when you drive over cat's eyes in centre of the road, a warning that you are straying out of your lane, out of your groove, out of your flow.

Most runs started like this, I reminded myself. Hemorrhaging doubts of even finishing it, berating myself for not going faster or farther. The negative chatter had the volume turned right up.

My mind escaped to a recent article that I had read about the benefits of being barefoot which claimed that standing barefoot on the soil and holding seeds in your mouth before planting enriched the qualities and quantities of the crop grown.

I summoned up recollections of time I had spent being barefoot in the past. Why was all this demanding attention now?

I needed another challenge - a focus. A sort of psychological sheepdog to round up my thoughts and move them on to new pastures. It had been a tough couple of years. I had spent most of it sat down on the spectator's bench, unmotivated and relatively uninterested in the game of life. Maybe it was time to join in again.

My trainers slapped the tarmac, unyielding and uncaring, petulance in every stride. Another previous conversation poked my consciousness.

A yoga teacher had questioned my need to run. "I don't think it's natural. We don't need to chase and catch our food any longer. It upsets me to see all these people out pounding the earth with their deadened feet. Listen to your steps, how loud and harsh they are. We all need to walk softly and gently over the earth, respecting her."

It had made me think - just not enough to stop running. Deadened feet. That phrase made me uncomfortable. I had tried to run more lightly but the thrill of striding out pacified my ego far more than treading lightly satisfied my conscience.

My mind writhed and wrestled with my thoughts. What was going on? Why the angst? Why the unrest?

I could hear 'challenge, challenge, challenge' in every footfall; an internal chant. I allowed the thought some creative space.

There were endless runs to choose from overseas, from the Great Wall of China Marathon to the popular pilgrimage, Camino de Santiago. I could maybe do one barefoot, seeing as that aspect had come to mind. But then there was the travel, the time and the costs. The practical aspects made me sigh.

I wanted to stay closer to home.

I could run around Wales; there was a new coastal path which was attractive and 870 miles long. I wanted to be the first to achieve something; there aren't many firsts left. Running the coastal path had been done. That distance would also be a hefty time commitment on top of an already solid work load.

So a shorter route then.

Across Wales maybe.

Barefoot.

Surely nobody had been crazy enough to have done that.

Discontent yielded under the enthusiasm of possibility. I continued to run wrapped in contentment, as if I had appeased some great energy by my awareness and consideration of the message given.

The seeds of an idea had been sown.

By the following morning, my idea was already in full leaf.

"I'm going to run across the width of Wales," I mentioned casually to my brother. "Barefoot."

"I'll be the first person to do it," I added quickly, by way of justification.

"No shit," he replied.

He straightened up, pushing his hands in the small of his back to do so and turned to face me.

"There will be a reason for that."

2.
Stepping into the unknown

'I am used to my head calling the shots;
it's a demanding task master,
we wrestle about a lot of things.'

My barefoot path began in the December of 2010. I was 43.

Whilst on a rare night out with a group of girl friends, I invited suggestions for my next personal challenge.

To date I had completed a parachute jump, several marathons, including a solo 4x4 expedition to Iceland to run the Reykjavik Marathon, a wing walk, and compiled and featured in several calendars, all to raise money for and awareness of, various charities.

I also like to have a goal to work toward.

When I was training for my motorbike license, the instructor warned me that my tendency to look down at the road surface just yards in front of the front wheel would not end well.

"Where you focus is where you'll end up. Focus on the pothole or the slippery drain cover and that's where you'll ride. You must look ahead, confidently focusing on where you want to go. When you go into a bend you focus on your exit from it."

It's the same when riding horses. When taking a horse over a jump I was taught to throw your heart over first. Keep your focus forward, on the path and direction you intend to take. Your intention leads the way, like an invisible trail of breadcrumbs.

Having a goal works the same way for me. It keeps me focused, on track, on my toes. Looking ahead.

"Another marathon" was the most popular answer from the girls, with various cities, countries and severities being forwarded. Then Adele piped up, "What about doing one barefoot. Or maybe just a 10K run - barefoot?"

As everyone else shrieked, "Ow. No. Impossible. No; something easier, more do-able," I was already kicking off my boots under the table.

Back at home, in the peace and stillness of my remote little cottage I was still mulling the idea over. I did what I always do when I am having difficulty making a decision. I made a list of the pros and cons:

The Cons - the Top Ten reasons I wouldn't/couldn't possibly go barefoot or 'unshodden' read:

1. I *love* boots; all types of boots – be they steel toe capped, high heeled, knee length, thigh length, ankle, fur lined, neoprene, lace up, riding boots or even waders.
2. I suffer from cold feet. In the winter I wear Sealskinz waterproof socks inside all of my boots. I often wear socks in bed. My feet are like blocks of ice from October to April. Chilblains are as certain as Christmas.
3. I'm a Piscean; Pisces' weakness is their feet. They need to take good care of their feet; they have to protect their feet.
4. I walk Yogi, my dog - a lot. The rougher the terrain, the more we like it; our preference to avoid other walkers means we hike 'the paths less travelled'; the more remote and rugged the better.
5. I am a landscape gardener. I wear steel toe cap boots all day every day; partly because they are practical and very comfortable but mainly because they are protective. They ensure my feet don't get mashed, mown, strimmed or spiked.
6. I love dry stone walling and am in the middle of training for professional qualifications. Steel toe capped boots are essential for the reasons above.

7. I am a bee-keeper. I wear wellington boots as part of my protective gear when I check the hives, as taught by my fabulous bee mentor.
8. I live on a tiny-holding (smaller than a small-holding) and every morning and evening feed, water and muck out my pigs and chickens. I live in Wales. There is a lot of mud.
9. I run a lot and have entered several 'races' for charity in 2011. I spend the same sort of quality time with my trainers as one spends only with best friends. I have several pairs. I love them all. Pulling on my trainers takes me into a parallel universe where all is well.
10. I am a public speaker. I speak in public. I have an exciting list of engagements for 2011. In public.

The Pros - the one and only reason I will give it a go

1. It appeals to me as a challenge. Simple as that.

I like the thought of the adventure, the journey and the experience. It already strikes me as being a 'heart-led' journey rather than 'head-led' one. Heart is an anagram of earth - I already work with the earth and like the idea of my feet, as well as my hands, being in direct contact with it. It feels primal, authentic and 'real'. It excites me.

But will my feet share my enthusiasm?

Perhaps they'll hate it. Perhaps I'll hate it. I'm used to the protection that strong steel-toe capped, leather boots offer. They allow me to indulge in my favourite things. I love my boots. All of them.

I am used to my head calling the shots; it's a demanding task master, we wrestle about a lot of things. What will happen if I shift the focus to my feet? What if I let myself be led by my feet? That already sounds weird. What if I take my mind off my mind? What if I let my mind wander? Will it come back?

3.
Walking the black dog

'Basically, she was considered to be a danger either to herself or others.'

I was ten years old when my mother was diagnosed with manic depression.

It has since become renamed bi-polar disorder which was deemed to be a more politically correct term; the condition itself is no more politically correct than it was as manic depression.

Of course she didn't suddenly become a manic depressive when I was ten. That's when she was diagnosed.

And like anything requiring and achieving a diagnosis, it already has a history.

All the rather unconventional and unusual behaviour that my brother and I grew up with gradually made a little more sense.

Or less sense. I'm still not sure.

Mum's diagnosis was made as a result of her being sectioned.

Being sectioned means "being admitted to hospital whether or not you agree to it; the legal authority for your admission to hospital comes from the Mental Health Act rather than from your consent." This is usually because you are unwilling to consent which in Mum's case was a symptom or condition of the illness. The term 'sectioned' just means using a 'section' or paragraph from the Mental Health Act as the authority for your detention. Subsequently we were told that a better word than sectioned is 'detained'. Mum would be detained under the Mental Health Act.

A better word maybe, but still the same heart wrenching situation; for us all.

The particular 'section' that Mum would be detained under, referred to her "having a mental illness which needed assessment or treatment and which was sufficiently serious that it was necessary for her own safety, or for the protection of other people."

Basically, she was considered to be a danger either to herself or others.

The first time she was sectioned was because she refused to come down from a tree.

She had been sitting in a rather majestic oak tree at the top of our garden for two days. Perhaps more alarmingly, we hadn't been the slightest bit concerned for the first 24 hours.

It was not abnormal behaviour.

For as long as I could recall, Mum had lurched between being the best Mum in the world and a woman I didn't recognise. When well, she was kind, caring and extremely committed to her family and home. Sometimes she would even stay up all night cooking and cleaning. Sometimes my brother and I would come downstairs first thing in the morning to find the kitchen floor covered in numerous piles of clothes and the twin tub churning away contentedly. The front room floor would be covered in newspaper with all the brass having been taken down and spread on the newspaper awaiting yet another clean and polish. And Mum would be out working in the garden.

She had endless energy.

In fact what she had was manic depression.

Welcome to the manic phase. Beautifully gift wrapped in the guise of hard work, dedication, commitment and caring for the family.

So cleverly disguised.

So misleading.

It is a cunning and manipulative illness.

These seemingly productive episodes would only be the start. They would quickly whip themselves up into more frantic and less acceptable incidents as layers of decoying behaviour fell away like the wrapping paper around the present in a pass- the-parcel game.

Blaring music, erratic driving, over spending, an unhealthy interest in other people's lives, endless consultations with fortune tellers and mediums, eye-wateringly high phone bills and other indiscretions would cause bitter arguments between her and my father. Occasionally friends and innocent, unsuspecting people would inadvertently be drawn into Mum's manic stage.

My brother and I would be ushered just up the road to my grandparents as one or both of my parents disappeared in fury into the night. My mother driven by adrenalin, my father by angst.

They were chaotic times hemorrhaging uncertainty and upheaval.

And then, as sure as dark nights followed dark days, depression took advantage of exhaustion.

Mum would tire and her mood would plummet. As hard as the manic bouts were to deal with, the depression was the most painful. For all of the family. Inevitably sectioned and hospitalised she would slide into a downward spiral from high energy into lethargy and exhaustion, refusing to get out of bed, eat or converse with anyone else in the hospital.

Back at home, clearing up followed the chaos.

The fallout and consequences of her illness-induced actions would be mopped up and absorbed by friends, neighbours and family. We lived in a small Welsh community. It had its advantages and disadvantages. The disadvantages included everyone knowing everything there was to know; the advantage was they were mostly silently supportive.

My brother and I learned not to question anything. It was adult stuff. There would be endless whispering and conversations which stopped when we came into a room but that suited us too. The often dire consequences were being discussed and dealt with on a need to know basis and we didn't need to know.

4.
The face (and feet) for radio

'I looked forward to the show but I dreaded having to explain my 'nakedness' to security at BBC Wales en route through the building.'

If I was going to raise money for charity by running a barefoot 6K, I needed to know more. Research on the internet immediately revealed there was an organised barefoot 6K being held in Battersea Park the following September, just nine months away. I am a great believer that if something is easy, or flows, it is a sure sign you are on the right path.

The barefoot run was in aid of Trees for Cities. The nature connection was surely an extra sign that this was meant to be.

I signed up.

Then I wondered - how the hell do you train for a barefoot run?

The distance wasn't an issue as I was already randomly loping around the odd charity 10K when time and incentive allowed. And of course my job kept me at a reasonable level of fitness. However it also kept me in steel toe capped boots.

Attempting to get my bare feet race fit would mean that I had to take every opportunity I could to go barefoot.

That would mean swerving the shoes for the (very) occasional night out, the even more occasional food shop, driving, walking my dog, giving talks, my radio work and other public appearances.

I had absolutely no hesitation about daring to bare my feet, the novelty even appealed to me. My reservations were all about people's reactions. The fear of being ridiculed and rejected far outweighed any physical fear.

Ego was alive and kicking - even without shoes.

The title and an icon 'Unshodden' were added to my website and I started to blog about my new journey. I also wrote about it in my weekly newspaper columns and talked at length about it on various radio appearances. Listeners and readers were used to my various charity challenges and interest was light. It was winter and Christmas was coming. Festivities versus bare feet? Festivities won hands down.

My very first unshodden outing was, rather appropriately, to meet the same group of girls who had instigated the barefoot journey just weeks earlier. During the day I toyed with the idea of turning up bare foot. It was still light and reasonably mild and seemed like an obvious first step. By the time I was ready to leave the house, not only was the truck windscreen starting to freeze but it was also extremely dark, the real black dark that you get when you live on a mountain. Suddenly it wasn't quite such an obvious thing to do.

So I did it anyway.

There is something very surreal about getting dressed to go out and then putting on a coat, hat and gloves. But not shoes. Or even socks. My feet were totally bare. It felt very strange and not surprisingly, I felt quite vulnerable. I was aware of a very loud voice in my head, "You can't go out without shoes on."

After negotiating the stony little lane from the cottage to the truck, I spent a couple of minutes sitting in the truck waiting for the windscreen to defrost, debating whether or not a freezing cold January night was really the right time to free my feet in public for the first time. I already felt ridiculous. Not brave. Not excited. Not curious. Just ridiculous. I wasn't sure I could do it.

I drove out of my narrow little lane with the bare toes of my right foot curled around the top of the accelerator pedal and my left foot trying to get comfortable amongst the grit and gravel that collects in the foot well of a gardeners truck.

As I drove, I realised I hadn't even put a pair of boots in the truck, just in case.

The car park of 'The Dragonfly' on the outskirts of Merthyr Tydfil was unexpectedly full.

I parked thirty yards away from the restaurant amongst heaps of dirty snow still piled up in some of the adjacent car park bays and I had no shoes or socks on. People were walking past my truck, wrapped up in coats, hats, gloves and scarves and furry winter boots.

My fear wasn't of any possible physical discomfort or danger but of being confronted, challenged or even chastised. I really didn't want to be asked to leave the restaurant. Not in public. I could tolerate cold feet but not the cold shoulder.

I had deliberately put on my flariest legged jeans in an attempt to hide my naked feet. I had naively hoped it would at least look as though I had sandals on – like that would be acceptable in December - but it didn't. It looked like I had forgotten to put my shoes and socks on.

I strode as boldly as I could manage through the gravelly car park - as much to cope with the discomfort of the feeling stupid as well as the cold - over the smooth cobbled entrance and into the carpeted foyer of the restaurant.

So far, so good.

The girls were at the farthest side of the dining room, so I simply focused on them, walked straight past the bar and along the full length of the busy dining room. We did the whole hug and greeting thing and I sat down, glad to be tucking my feet under the table.

No one, not even the girls, had appeared to notice my semi naked state.

No one had asked me to leave.

No one cared.

I quickly confessed my barefoot state and was hugely supported and encouraged by these special women. A few people looked over at the cheers and clapping but remained completely uninterested in my bare feet. I was incredibly relieved.

Throughout the evening, my feet remained unremarkable and under the safety and security of the table. I was constantly aware of their nakedness. I felt every little change in temperature, the soft carpet beneath my toes, the hard wooden cross bar of the table, the draft every time the door opened; they seemed unperturbed, happy to be bare, and contented somehow.

And so was I.

Just a couple of hours later, the return trip through the car park to the truck seemed far more natural and almost normal. Fear and anticipation had already morphed into excitement and achievement. My feet were tingly; cold but also alert and aware.

Having to be constantly aware of my surroundings had also kept me very much more focused on the present moment, in the now. My feet felt pleasantly energized and alive.

And so did the rest of me.

And as I lay in bed, replaying the whole unshodden experience, I was still very aware of my feet. Despite being exposed to grit, puddles and below-freezing temperatures, they felt warm, enlivened and just connected to something powerful or even everything.

And so did I.

That first outing with naked feet had left me feeling as though I was walking on air and the next stage was to talk about it all on air.

Radio Wales had always been wonderfully supportive of my various fund raising antics and bran-tub lifestyle and so Jason Mohammed extended an invitation to his lunchtime show to discuss my unconventional bare foot path.

I looked forward to the show but I dreaded having to explain my 'nakedness' to security at BBC Wales en route through the building. I wondered if they would insist that I wore shoes to the studio. I hoped not.

I needn't have worried; they were all absolutely fabulous. As two other visitors felt it necessary to point out that I was barefoot, John the security guard simply shot them a look which I suspect made them feel the odd ones out.

John was cool and completely ignored my exposed toes, instead asking after Yogi, my dog who he had seen in the truck with me via the CCTV. He shared stories about his dippy Doberman and feisty terrier and also shared his love for a rather unconventional zombie garden sculpture which he showed me on his mobile phone.

Bare feet weren't going to faze this chap.

The phone-in production team were far more intrigued and questioned me as I waited to go into the studio.

Buoyed by their enthusiasm, I joined Jason in the studio, put my headphones on and, live on air, Jason shared my whole unshodden status with the listeners.

There was no going back now.

As direct as ever, he launched into it, "Lynne, all the email interaction we have had you have always signed off with a very driven, "Boots on ...". Today you tip up at the studio without any shoes and socks on. What's it all about?"

I explained as best I could. It sounded a little tenuous and vague but I felt very grounded and authentic. It was a new experience, a new challenge. I didn't have a lot to share just yet, apart from my enthusiasm and anticipation.

And it seemed that was enough. Listeners were intrigued. Festivities were over and forgotten. It was the New Year, a time of new resolutions and challenges. My fellow guest on the phone-in panel, the Deputy Chairman of Glamorgan Cricket Club, was suitably suited, booted and was not only impressed with my unshodden status but also complimentary about my feet!

So it transpired, not only do I have the face for radio but also the feet.

5.
Sevens and eights

'Dad would just look straight ahead.
I tried not to look at the distorted faces pushed against the bars but didn't have Dad's resolve;'

By the time I started high school, Mum had become a regular patient at Mid-Wales Hospital, originally the Brecon and Radnor Joint Counties Lunatic Asylum, in Talgarth, Wales. The psychiatric hospital was opened, amid public ceremony, on March the 18th 1903, by the Rt. Hon. Lord Glanusk who said of it: "everything has been done that human ingenuity could devise for the happiness and safety of the inmates, and under the blessing of God, for their speedy restoration to health."

March the 18th is my birthday.

Such was the notoriety of the hospital, it was simply known as Talgarth locally and to those who had any connection to it. Like a celebrity being known by their first name only. Lulu, Cher, Madonna, Talgarth. Talgarth is actually the hamlet but whenever we explained 'Mum is in Talgarth', people knew she wasn't visiting the corner shop.

With hindsight, we were incredibly lucky that the hospital was so local. Although lucky was not an adjective that sprung to mind as a teenager visiting her mother in a mental hospital. They were no longer asylums but they were still mental hospitals, or loony bins or nut houses according to the tireless taunting of several mean school children.

Despite being local, the round trip to Talgarth was thirty miles. Thirty miles of narrow country lanes. Thirty long miles, winding roads, often undertaken in the dark, the wind and the rain.

Dad made the thirty mile round trip every single night that Mum was in hospital. It became a sort of pilgrimage for my father; the operative part of that word being 'grim'.

I often accompanied him; homework having to wait until I got home later. My brother was considered too young to visit regularly but not too young to stay at home on his own. It must have been a tough decision for Dad and no doubt triggered 'exclusion issues' for my brother. Of course he was still included in weekend visits. They were the closest we got to family days out. We have since discussed those visits and he has very little recollection of them. He's lucky.

I still feel a lurch in the pit of my stomach as I recall the left hand turn through the huge wrought iron gates in to the hospital. The gates alone were like something out of a horror movie.

As frightening as the trip would often be, it couldn't even begin to compare to the absolute fear that the time inside the hospital building evoked.

Mum was on a second floor Ward called East 8, or '8's'. There were two entrances. The main entrance that led straight up the stairs would often be locked after dark. Dad and I would have to use the other entrance. That took us through the downstairs ward called 'Sevens'. Ward East 7 comprised of the padded cells. Patients would push their faces up against the bars on the door, foaming at the mouth and letting out blood curdling screams. They would leap around their cell, rebounding off the walls like monkeys at a zoo.

It was terrifying.

I would walk as close to my father as I possibly could, often bumping into him as I glanced behind to make sure we weren't being followed.

Dad would just look straight ahead. I tried not to look at the distorted faces pushed against the bars but didn't have Dad's resolve; the animalistic howls would command my attention.

And I knew we had to come back the same way.

Mum's first hospital admission stretched into a long and difficult eight months. They had to find the appropriate medication. It turned out that Mum was particularly sensitive to medication and as well as horrific physical side effects, her mood continued to swing from exhaustive highs to crippling lows. She would often be on suicide watch with a nurse sitting patiently outside her door reading a paperback book.

When she was 'high', Dad and I would sit in the day room with the other patients and visitors as Mum sped around completely uninterested in our visit. Dad was always great, chatting to other patients often about utter nonsense. I admired him so much. He made the most of his cup of tea and a sit down.

I always refused a drink; I was scared of catching madness.

The highs were tiring and even ever so occasionally entertaining but the lows were just excruciatingly painful. Shrouded in depression, Mum would curl up in her bed like a wounded animal, refusing to engage or even acknowledge our presence. Her eyes would be devoid of emotion as she just stared ahead. Knowing she was awake was even worse. If her eyes had been closed I could have pretended she was asleep. There was no such relief. No escaping the darkness.

No words.

Dad and I would leave in silence. Walking back through the howling bedlam of Sevens, back to the car park, back along the narrow twisting lanes, back to the house, back to my brother, in complete silence.

Still no words.

Silence became an ally; I learned to keep quiet, not to ask questions, not to risk upsetting anyone or anything, just to maintain the silence.

Silence was safe.

I also learned that sharing anything emotional with Mum was dangerous. Mania doesn't even tip its hat to confidentiality, let alone respect it. During her highs, conversations shared in confidence either became problems to be fixed or even worse, the subject of even more manic merriment. As a teenager, that was excruciatingly painful.

I learned to cook, I learned to organise and I learned that the only certainty in life was uncertainty.

6.
Going swimmingly

'"Don't see the point of it myself," he grinned nodding toward my naked feet. Ever enigmatic, he added, "Impressive though."'

'There were an estimated 25,700 excess winter deaths in England and Wales in 2010/11. The excess winter mortality index for England and Wales was highest in Wales; in the two previous winters it was highest in the South East of England.' It is a rather random statistic that found its way into my consciousness as I stubbornly continued to practice my barefooting in the harsh winter conditions. December 2010 turned out to be the coldest December since Met Office records began in 1910 and I was officially, and delighted to be, 'snowed in' on my beautiful Welsh hillside several times during the winter months.

Of course I had to try walking in the snow barefoot. It wasn't my finest moment. The snow surface had frozen and it was slippery and surprisingly sharp. It wasn't long after the initial thrill and excitement, that I 'cold-footed' it back indoors. That box had been ticked.

And anyway, I wasn't obligated or even tempted to be bare foot all the time. That wasn't the deal. I was still donning my steel toe capped boots for my landscape gardening work, for woodland walks with Yogi and when I was rushing around devouring day to day chores. Haste and bare feet do not make good bed fellows. However, I continued to spend as much time as 'safely possible' barefoot, raising eyebrows and awareness of my charity challenge that would be taken later in the year.

I also continued to fund raise for other charities in the meantime and launched Marie Curie's Swimathon event in Cardiff with Duncan Goodhew in the spring of 2011.

I had worked with Duncan before. He is a lovely, kind, gentle man who always puts me at my ease.

Of course we were both barefoot at the pool side.

I realized that the support of celebrities would only smooth my new path, so contacted author and Taoist, The Barefoot Doctor. I had read some of his books. I liked what he had written and decided I would like to meet him. I could only ask.

He proved to be incredibly approachable and supportive and we met for lunch in Crickhowell. I was a little taken aback to see him clad in socks and Crocs but he explained that the whole Barefoot Doctor title referred to a basic approach to his holistic practices rather than actually being 'sans shoes and socks'.

"Don't see the point of it myself," he grinned, nodding toward my naked feet. Ever enigmatic, he added, "Impressive though."

With time flying by and temperatures dropping just as quickly, I decided to use a gym to maintain my training schedule. It would mean I could still train on those really awful harsh wintery days. A local gym politely refused to allow me to train bare foot citing health and safety issues. But the wonderfully accommodating management team at The Manor Hotel in Crickhowell, allowed me to use their gym in the off-peak periods. Without trainers. And without hassle.

The soles of my feet were becoming more used to their increased exposure and were toughening up nicely. Despite having become used to running on tarmac without trainers, running barefoot on the running machine was quite a different experience. The grooved rubber running belt was unforgiving and uncomfortable. I also sounded extremely heavy-footed and my calves ached like hell. My legs felt like they belonged to someone else.

I wasn't that bothered as it was still only January and I had 8 months left to train. Eight months to be able to run six kilometres. Bare foot. Surely that was do-able.

I hoped morale would improve along with the weather. I was still working outdoors, all day, most days and it was hard going.

I had been tearful whilst working in freezing temperatures and not even being able to feel my feet despite them being safely wrapped up in two pairs of socks and neoprene-lined boots. The cold can be cruel; it was also exhausting.

The stubborn streak that flows though me ensures the harder it is to do something, the stronger my desire to do it. The steeper the climb, the stronger my desire to enjoy the view from the top.

I had also been doing a little barefoot ground work in order to give my feet a boost along on this barefoot path. I had researched barefoot shoes (a strange contradiction in terms, referring to running shoes that have none of the padding, support and cushioning of conventional trainers.) and decided to try the Vibram Fivefingers - a kind of foot glove.

I also bought a cheap pair of ‘folding shoes’ which I could just roll up and keep in my pocket in case I encountered a hostile authoritarian or environment.

They were little things that made a big difference to my confidence and comfort.

Spending time barefoot was proving to have far wider connotations than becoming ‘hard-soled’. I was definitely 'finding my feet' in other areas of my life. Ditching the footwear was like dropping a mask. Despite feeling more vulnerable, I was already feeling more comfortable in my skin. As the skin on the bottom of my feet became thicker, so did the rest of my skin. Being barefoot suited me.

My interest in self-improvement and personal accountability had introduced me to the concept and value of 'stepping up' in one’s life; I just never thought I'd be taking my shoes and socks off to do it.

I am a big advocate of the powers of visualisation, manifestation and cosmic ordering, or asking as I prefer to call it. It gives me a lot of strength and reassurance. My first unshodden trip to a supermarket was an ideal opportunity to 'create the outcome I wanted beforehand'.

During the drive, I pictured myself walking softly along the supermarket aisles, choosing the food I wanted, and going through the checkout quickly and quietly - all without shoes and a showdown.

And that's precisely what happened.

At the busy entrance of the supermarket a woman kindly passed me her redundant trolley and looked directly at my feet – perhaps that's why she gave me the trolley – but nothing was said.

Once through the doors, I chose a quiet aisle to give myself time to adjust to my environment. Gathering courage I wandered over the cold tiled floor to the fruit and veg and the luxury of a rather randomly placed mat.

I asked a guy who was replenishing the banana box, if he would mind taking a photo of me helping myself to carrots.

"No problem," he said, taking my phone, "do you want all of you - like your feet as well?"

Ha! He'd noticed.

"Yes please, particularly my feet," I smiled, throwing down the gauntlet. Inviting a response. Getting just a little cocky.

"OK," he said and took a couple of photos. As he handed me my phone back he said casually, "That's probably the maddest thing I've been asked to do since I've been here." And went back to his bananas.

And that was it! No questions, no queries, no chastising, no manager called, no problem!

The rest of the grocery-gathering was equally uneventful. I made my way to the till, packed, paid and left. Unshodden and unchallenged.

Once back at the truck Yogi's cheerful little face inspired me to tiptoe over to Pets at Home to do her shopping. Hemorrhaging bravado, I bought her dog treats and asked the guy on the checkout to take another photo for me.

"No problem," he said, "but you're not from Health and Safety are you? They wouldn't like this."

"Do I look Healthy and Safe?" I laughed. "I've just been into ASDA," I challenged. "They didn't mind."

"They wouldn't care if you were naked," he retorted. "But you won't get into Tesco. I used to work for them and they're mad keen about Health and Safety; you'll never get into Tesco without shoes on."

7.
Putting the 'rat' in g-rat-itude

'As a result, I spent more and more time in the Headmasters office. It became a situation that suited him far more than it suited me.'

Having a mother that was in a mental hospital as I started high school was not the best way to ingratiate myself into a pack of unfamiliar, unrestrained teenagers.

Unsurprisingly, I proved to be nothing short of 'bully fodder'. Word spread quickly 'Lynne's Mother is a nutter' and so did the pain, both physically and emotionally. The physical aspect was bearable. Just. Emotionally, I found it all crippling.

I got used to the hair pulling and being pushed from one laughing bully to another, the name calling, the contents of my school satchel being tipped out. The school bus was the worst. There was nowhere to go, nowhere to hide. I would get on the bus and think, "not long now until it all starts." The bus journey was a tortuous 45 minutes. It's a long time when you're being bullied.

By the time I got to school I would be in no mood for lessons or sarcastic teachers, so I would slope off to find some peace and quiet. It was called bunking off or skiving. I was just trying to avoid the cruel circus that school had become.

I earned the reputation of being a difficult child. I was just a child in a difficult situation.

As a result, I spent more and more time in the Headmaster's office. It became a situation that suited him far more than suited me. But by that time, it was the lesser of several evils. And his office and leery behaviour was the lesser of those evils.

I hated being at home and hated being in school. I didn't feel comfortable, wanted or safe at either. And worse of all I was convinced it was all my own fault.

At the age of 14 my only sanctuary was the great outdoors. The space and peace of the countryside and the unconditional love and affection of the animals eased the pain that humans inflicted.

Animals were always pleased to see me, never hurt me and everything was OK whenever I was with them. They were very soothing and undemanding; there was no drama. They just got on with things in a beautifully uncomplicated way, unconditionally.

Environments that involved people were, in complete contrast, angry, volatile, uncertain, dangerous and fearful spaces. I had to be on my guard. Constantly alert. My head would ache. I hated it all.

I swiftly built up a dislike and distrust of people. Like a wall of stone, I built it to protect myself. I built it to hide behind and I built it to last. Animals took the place of human friends and the higher up a mountain or deeper into a wood I could get the happier and more relaxed I became. That, of course, fuelled the bullying; not only was my mother a 'nutter' but I was a freak who spent all her time in the woods talking to animals.

My animal allies worked in strange ways. I started taking my pet rat to school with me. It was a defiant act that was to provide the most delicious form of defence.

I would take him down my school shirt. It seemed completely rational even if a little rebellious.

And it had the completely unexpected result of stopping any unwanted attention from inappropriate sources. I don't know who had been the most surprised.

8.
Dai The Foot

'Can't see how that would work –
are you sure it wasn't just a kinky thing?'

I am a practical person. Therefore as I continued to become 'barefoot fit' for my unshodden 6K in just a few months, it made sense to go and see a chiropodist. I had reached 43 years old without having to visit one so it seemed as good a time as any. I made an appointment. I did consider walking into his 'surgery' barefooted but it seemed a little bit too forward. After all, I wanted his advice.

I would tread lightly and just bring it up casually in conversation. I was keen to get a 'professional' view on my quest and I presumed that as a professional he would say 'don't do it.'

To the contrary.

David Richards is better known locally as 'Dai the Foot'. He is quite simply one of the nicest people you will ever meet. My brother was already a regular client but this was my first visit.

On my arrival he immediately asked after my brother's dog, Diesel. Not my brother but his dog!

"He usually comes to me after work so brings Diesel with him," he offered by way of explanation. "Lovely dog."

He checked out the little indent on the sole of my left foot which he rather alarmingly diagnosed as a verruca.

"Nothing to worry about at all, far too much fuss made about them these days. We all have memories of feeling like the great unwashed if we had a verruca in school - banned from the swimming pool or changing rooms - silly." He tutted. "It should clear up on its own, clever thing the verruca virus; it can disguise itself and creep into your cells without being detected. Then one day out of the blue the cells will rumble it; they'll spot it as an imposter and it's evicted. That's the end of it. It can't survive once it's been revealed as a baddie, so it just buggers off."

How cool. I was so impressed I nearly forgot to mention my challenge.

Suddenly, it was looking good; surely someone who likened my verruca to a conman would be less likely to berate my barefoot intentions?

So I told him.

He didn't even look up from filing my toenails.

"We are meant to be unshodden," he shrugged, "just not on tarmac, pavements and concrete. That's what makes it unnatural today. We are supposed to be unshodden but on grass, mud and peat - perfect for your job!"

There was a very real risk of me being rendered speechless as he continued. "People blame their shoes for all sorts of problems with their feet but unless the shoe is a poor fit, mostly they are problems that would occur even if they didn't wear shoes. Most problems like bunions are genetic and down to the make-up of your foot. You've got great feet. Bio-dynamically they're strong and good. You'll have no problems being barefoot."

"Oh? Really?"

"Just don't visit any Breaker's yards," he added.

I really liked this guy.

He told me that a little bit of banana skin put over the verruca (inside of banana skin to skin on foot) and held in place with a piece of duct tape often gets rid of verrucas too.

I reciprocated by telling him that I'd read that the Romans used to beat the bottoms of their feet with bunches of stinging nettles to harden the skin.

"Really? Strange ones the Romans," he replied. "Can't see how that would work - are you sure it wasn't just a kinky thing? This is interesting, you're making me recall stuff I learnt as a student now; the skin is three or four times thicker on the soles of our feet anyway and it's possible to callous it by exposing it and walking barefoot on different surfaces. On a healthy foot the pressure will be evenly distributed which means the skin will thicken up uniformly. Problems occur when ill-fitting foot wear means you get a concentrated pressure point and hard skin forms as a corn. You shouldn't have any problem with your feet at all."

I share another one of my researched facts, that we use 4% less oxygen when running barefoot. He was far less impressed than I am.

"Let me know how you get on. Oh," he added with a wink, "and tell me if the nettles work."

9.
Horses for courses

'She had seen me on TV leading up one of John's horses at the 2.30 at Chepstow Racecourse.'

Every bully has a group of followers; those who are scared of being bullied themselves if they refuse to join the instigators. But there is always one main bully, one ring leader. And the girl who happened to be the main instigator of my bullying in school also lived in the same small hamlet as I did.

I was frustrated to discover that she was actually quite sociable and friendly when our paths crossed with no onlookers and then confused to be confronted with her more familiar aggression and poison when we were back in school.

Confronting her obviously wasn't an option so I put up with it. I was used to keeping quiet about my home life; silence was golden. It wasn't my smartest move; because of the 'perceived friendship' out of school, no one realised the reality or severity of the bullying during school time.

As she came to the house when we were 'friends', she was also able to gather ammunition from my mother's manic antics. Ammunition which was used to maximum effect when Mum was next sectioned.

It was horrendous. And incredibly unjust.

So instead of just missing lessons, I started to miss the bus too. I would skive off school altogether.

It wasn't difficult. I had to get on the bus about half a mile from home and then the bus would stop for the next pick up 4 miles further on at the next village. I would simply get off and walk very slowly home. I would stop and talk to the ponies en route or just sit under a tree or in a barn if it was wet and cold.

Pretending to be ill was pointless. "You're going to school unless your head is hanging off" was the rule of thumb in our house.

Of course one of the ironies of Mum's illness was that it was easier to avoid school than it would have been had she been well and even vaguely concerned.

The highlight of my mornings as I got ready for school that I wasn't going to, was watching a string of locally trained racehorses trot up past the house. I would wait and watch them from my bedroom window; waiting to hear the heart-lifting clip clop trot of the string of nine or ten thoroughbreds. They would stream past under my window, the horses blowing, the orange rugs flapping under the little racing saddles and the jockeys standing high in their short stirrups crouched low over the horses necks. I loved it.

The National Hunt stables were down in the village and I was in awe of both the magnificence of the horses and courage of the people that rode them. John Bowles was the trainer. He would always lead the string of horses and jockeys. A strong, charismatic figure, people talked about him with either adoration or with hatred but they always talked about him.

He soon noticed that I was not at school when I should have been and often spotted me stroking the horses in the fields. I would straighten a rug that had slipped, re-tie a loosened hay net or just stroke and chat to the most magnificent animals I had ever seen. He suggested that I went and helped on the yard, gruffly suggesting that I 'make myself useful'. And so my web of deceit grew larger as I would pretend to go to school and instead wander down to the yard to help with the horses. I would pass his own kids as they went to the same school as I was supposed to be at. His daughter was in the same year as me and his son in my brother's class.

No one said anything. They were rules I understood. It was fabulous. I felt safe.

Then one day I went home after an 'alleged' day at school to find Mum particularly agitated. She had seen me on TV leading up one of John's horses at the 2.30 at Chepstow Racecourse.

After the initial drama, interest in my chosen activity soon waned so I continued my yard duties and eventually got to ride out the strings of horses in the early mornings. I was 14. I would pass the house and wave to Mum and often pass Dad in work too.

It had been accepted that as I obviously had no intention of going to school, it was better to know where I was.

I would return to school periodically, usually after a threatening letter had been written to my parents. I would soon tire of the inane taunting and having my satchel emptied, and return to the stables.

I loved it. I was treated well. I had responsibilities. I worked hard. I got tired. I slept well. I felt I belonged somewhere. I was praised and appreciated. At last.

My respect for the horses grew and grew; I was in awe of their nobility and power, their gentleness and strength. My regard for humans waned; I despised their shabby, disloyal, lying, cheating behaviour.

It transpired that human behaviour on the stable yard was similar to what I was used to at home. There were lots of hushed conversations. Secrets.

I would frequently be told, "You haven't seen me, right?" or, "Don't tell anyone."

My mother's mantra had become don't tell your father. It was all so familiar. I was used to keeping secrets. It was fine with me. I didn't care what they did anyway. I just decided that I wanted to be around adults less and less.

Not only was I used to keeping quiet I was used to becoming invisible.

10.
Bees and celebrities

'Several celebrities also added their support'
'It is what it is.'

March came and went and with it, another birthday. My 44th. All the fours. Spiritually and numerologically, a 44th birthday is special, two fours make eight. Eight is an auspicious number. Maybe it would be an auspicious year but my barefoot birthday was pretty much the same as any other. Secretly (or not quite so secretly now), I enjoy my birthdays. I like reflecting over the previous twelve months and what they have taught. I also like planning goals for the next twelve months. This year, that was already in hand. I was already striding toward my barefoot 6K challenge in September.

I am reminded of the saying, "I try to take a day at a time but sometimes I get hit by a week at once", as the time was whistling by at an ever-increasing speed.

My barefoot journey was leading me to some pretty special people and I remain uplifted, inspired and at times a little overwhelmed by these wonderful gentle souls that wafted across my path.

I interviewed the inspirational Perry Haldenby for one of my newspaper columns.

A few years ago he had been a body builder who owned a couple of nightclubs and enjoyed late nights, red wine and red meat. He swapped all that for yoga and raw food. During our interview we discussed biodynamic gardening and universal energy. He shared many wise things which resonated, one of them being, "We're at where we're at. We must remember to trust in all things spiritual and to be true to ourselves."

It reminded me of a favourite mantra of mine which is, "It is what it is."

We're at where we're at.

It is what it is.

The power and profoundness of these lines are belied by their simplicity. Try it. That's all.

It is what it is.

Perry also shared that despite his path not being an easy one, it is nonetheless 'his choice' and that by managing his lifestyle and diet so consciously it resulted in him taking increased responsibility for other aspects of his life too.

I experienced the same emotions as I spent more time barefoot. Needing to be aware of where I placed my feet kept me organically focused and in the now, the present moment, and that elevated awareness and acceptance of personal responsibility naturally seeped into other areas of my life.

It felt good. Not always comfortable but good.

My own diet had also improved dramatically since spending time barefoot. It's as though the greater the connection that was established with the earth and with Mother Nature, the greater the desire became to eat and live in a way to honour her and, by default, myself. There was a greater internal connection too, as though all the dots began to join up and make sense. Clarity was improved, concentration was sharper and decisions were easier to make. I no longer wanted to eat meat. My body told me it no longer wanted to digest flesh. It was a decision made on a cellular level, not a conscious one. I just listened and honoured its advice. And it was remarkably easy. A lot easier than committing to decisions which were made with my head. I was authentically making healthy choices and adhering to them effortlessly.

I'm sure it was no coincidence that spending time barefoot was also leading to an increase in energy - both spiritually and physically. And of course, I wasn't barefoot all the time. Just by choice and when I deemed it safe to do so. To be barefoot I had to be hardy, but to be barefoot when using garden machinery for example, would have been fool-hardy.

And I still loved my boots - if I wasn't bare footed then I was fare booted. The best of both worlds.

But the further I travel along this unshodden path, the less I want to be shod; the more I want to be barefoot. It's addictive or maybe it's just authentic. Or could authenticity be addictive? Is it our natural state? Being barefoot and authentic? Maybe.

Stimulating and exciting projects continued to flow. Like attracts like. The Royal Horticultural Society asked me to create an inspirational and interactive Show Garden for their Spring Flower Show in Cardiff. They wanted to increase visitor numbers and wanted something a little unusual to attract both visitors and the press.

Building, displaying and promoting my Bee Friendly Garden as a Barefoot Beekeeper was interesting to say the least. Understandably, Health and Safety insisted I wore my steel toe caps as, together with my brother Ian, I built and dismantled the garden but the three Show days were spent 'sans shoes and socks' with the public showing genuine interest in both bees and being barefoot.

Purely for research reasons I had been tending my own hives barefoot for a fortnight before the show and hadn't been stung. It added to the intrigue and interest.

Several celebrities also added their support to the Bee Friendly Garden; Jason Mohammed brought his family, Wynne Evans (Gio Compario of Go Compare fame) became 'The Sting', donning a bee-keepers suit in a mickey-take of Top Gear's 'The Stig'. Jeremy Clarkson politely refused an invitation citing his lack of interest in both bee keeping and gardening.

Radio 2's traffic reporter Lynne Bowles was far more interested in the fact that running bare foot used 4% less oxygen than running with trainers than Dai the Foot had been and BBC Wales filmed me barefoot and bee-ing friendly for their weather reports.

Following the RHS Show, I was asked to present a talk to the local primary school children during their Eco-Week. I was thrilled when they welcomed me into Assembly and they had all taken their shoes and socks off in my honour. That was very special.

I spoke to 200 women at a Ladies Luncheon Club in Stratford upon Avon. They told me they would all be wearing hats to mark the Royal Wedding and asked that I wear one too. I readily agreed, reminding them that I would also be barefoot. That was a first - giving a talk in a big hat and bare feet. The talk was titled 'Celebrity Gardening Secrets' but as was becoming the norm, they were more interested in my feet than fuchsias.

But it was not all media and merriment. As well as keeping the business going, I was still 'in training' for my barefoot 6K which would take place in just three months.

I was taking a little barefoot walk most evenings, running in my Vibrams Five Fingers twice a week and running barefoot as often as I could in between.

I was also trying to keep my feet as dry as possible which meant hanging them over the edge of the bath instead of soaking them as I yearned to do. Running in the rain was proving to be unproductive as my feet started to 'shred' when wet and soft.

People were getting used to seeing me wandering around without shoes on locally and even Yogi got less embarrassed about the whole thing.

It made me far more aware of her little bare paws when encountering hostile surfaces. Tarmac gets very hot. We should be more aware of where we walk our four-legged friends at different times of the year and accommodate them accordingly. Something I would never have thought about unless I had shared their experience.

Something else I found was that walking downhill shoe-loose and fancy-free is far harder and more uncomfortable than walking uphill. As I walked downhill it felt as though the contents of my feet were sliding down into my toes; it was weird and comical at the same time. Walking downhill on tarmac also created more friction on the soles of my feet. When running, the effects were increased noticeably.

I was still using my Vibrams for some of my tarmac runs. I wasn't worried about building up to 6k barefoot on tarmac, as the run was in Battersea Park which would offer the luxury of running on grass for part of the course, if necessary. My path was proving to be far less about running barefoot and more about the unexpected benefits of spending time barefoot.

By the end of July and amid a busy summer, I was feeling more and more like the Mad Hatter from Alice in Wonderland as my life seemed to have become enveloped in an air of insanity.

11.
I am my father's daughter

'He wouldn't have won any Father of the Year awards but he did his best.'

When Dad was just twelve, his father died. His mother was a well-respected and much loved teacher and doted on my father and his brother, John. They had a poster of the world on their kitchen wall and whenever any country was mentioned on the radio the boys had to find it on the map and discuss any consequences of the story and the impact it may have on neighbouring countries. I was mightily impressed by this story and also a little irritated that he hadn't passed on such a clever learning opportunity for me and my brother.

After attending college, Dad had set up his own gardening business. By his own admission, he had been too young and it hadn't worked out so he went to work for BWD reinstating and landscaping the sides of motorways and roundabouts. We could never pass the Aust section of the M4 near the original Severn Bridge without Dad telling us he had seeded all the grassed areas there.

Whilst working on the roads in Surrey, he had met my mother. He persuaded her to move to Wales and, soon after, her parents followed and moved into a little cottage just half a mile up the road from Mum and Dad.

When my brother made an appearance two years after my own debut, my father decided that taking a job as a postman would give him the security he felt obligated to provide for his family and also allow him to hobble in the afternoons to keep his green thumbs happy.

He was a grafter. He would start work at the sorting office at 4am, finish his deliveries at 11am and then go on to undertake various gardening work until 5pm. He was also asleep in the chair by 7 o'clock and in bed by 8pm every night. Nobody was allowed to go upstairs after he had gone to bed for fear of waking him up, unless it was to go to bed ourselves.

He worked hard and those were the rules.

Each night we tiptoed upstairs in the dark and in fear of waking him.

He wouldn't have won any Father of the Year awards but he did his best. He was the best Dad I had and I know my brother feels the same.

He insisted we called him Bob. Everyone knew him as Bob, except my Mum's mother and my Uncle John who called him Rob. His own mother had been the only person who called him Robert.

When my brother and I wrote about 'What We Did In Our School Holidays' in primary school, the teachers thought Bob was our dog.

He didn't want to be called Dad. He told me he had never wanted children. I remember waiting for him to add that he was glad he had. Those words never came.

We weren't an emotional demonstrative family. I don't know if it was a generational thing or even a Welsh thing. Come to think of it, we weren't terribly well emotionally equipped, let alone demonstrative. It was deemed more important and productive to be able to work hard and just get on with it, rather than to talk about it

So we did.

12.
Going the whole hog

'It seems 'the chances' that the vet mentioned are just that – chances.'

2011 was also known as The Year of the Pig for me. It was in fact the Chinese Year of the Rabbit and it was those rabbit-renowned tendencies that made it memorable.

As well as the day to day running of the landscaping business, on site consultations, bee keeping, writing, speaking, radio work and training, I became a piggy midwife. The previous year, I had bought three little pet pigs. I had travelled to the Lake District to collect one but had been unhelpfully mindful of the nursery rhyme and had come home with - three little pigs.

They were the small variety; I refrain from calling them micro pigs as, in my opinion, if you want a pig that will fit in a tea cup, get yourself a guinea pig. However, they would still be manageable at the size of a small Labrador when fully grown.

As the result of numerous requests for little pigs from friends and colleagues, I made arrangements to put the smallest of the gilts to a little boar. In an unexpected sequence of events it turned out to be the little boar that Alex Reid gave to Katie Price at their wedding. Following their super-quick separation, the little boar had been returned to his original breeders several miles away from me. Complete with pink nail varnish on his trotters!

His dubious trotter adornment didn't put my girls off.

Despite my intentions to just mate one of my gilts, the boar proved to be far quicker than I was and helped himself to all three quicker than I could say 'nail varnish remover'.

I panicked and plucked up the courage to ask the vet if there was such a thing as the morning after pill for pigs. He assured me that as they were a) small b) first time mothers and c) likely to roll on one or two of their litter then the chances were I would only have maybe seven or eight piglets to nurse from all three pigs. It was better to let nature take its course.

There are seemingly endless fascinating facts about pigs, one of them being that their orgasm lasts around thirty minutes. I did the math; I had had at least an hour and a half to have prevented my piggin' predicament. Another fact is that a jet engine can reach 113 decibels on take-off, a pig can squeal at 115 decibels - another fact I had actual experience of.

And a pig's pregnancy lasts for three months, three weeks and three days.

As the farrowing time approached, I slept in the barn with the pigs, keen to avoid any squashed piglets. It was a harrowing farrowing for sure. Suffice to say, after a severe lack of sleep and superb midwifery skills, at the end of June 2011, I had twenty-seven healthy and happy piglets to look after. And three fraught sows.

It seems the chances that the vet mentioned are just that - chances.

Not surprisingly my barefoot endeavors took a step backwards as I fed, mucked out and generally tended thirty pigs after a hard day's work. There were pigs everywhere, every shed, building and even a borrowed horse box had been parked on the driveway to provide temporary housing.

I was almost too tired to put one foot in front of the other - shodden or otherwise.

I blogged:

I have lost my sense of humour, my mojo, my energy and my enthusiasm. I am tired. Painfully tired. I just want to sleep. For a long, long time.

Whilst I still had the fond memories of experimenting with care-free and shoe-free days, the security of my steel toe capped boots had proved pretty essential during that time.

My feet had been imprisoned again, denied their freedom. I felt the difference.

I reminded myself, "We're at where we're at. It is what it is."

All too soon, race day arrived.

I milled around nervously at the start of the barefoot 6K under a colourful banner announcing the '2011 Treeathlon for Trees for Cities', in Battersea Park. With the other runners, I went through warm up exercises, stretching and nodded at each other in recognition and anticipation. Some runners were wearing barefoot shoes, while others, like myself, were completely bare footed. I had my Vibrams in my hand. Some runners were looking extremely competitive. Others, like myself, looked rather anxious.

The park paths had been swept the previous day but high winds during the night had brought down prickly beech masts, spiky conker cases and gnarled little twigs. There was no respite on the grass verges, they too were strewn with nature's debris and were looking unhelpfully uninviting. A delay was announced over the tannoy and I reflected over my barefoot journey thus far.

It had been a tricky path to negotiate. Admittedly my feet had spent more time safely ensconced in boots than I had intended. I still had to pay the bills. I may have embraced the benefits of taking the time to be barefoot but my bank manager had yet to be convinced.

I had started the year well, despite the cold. My main challenge had always been the weather. Car parks had also been an unexpected obstacle. They have some of the dirtiest surfaces imaginable, often with a sprinkling of undesirable objects from broken glass to small tacks. Nonetheless, those and other obstacles were overcome with a healthy mix of enthusiasm and nervous energy. While visiting supermarkets, restaurants, giving talks and presentations, radio shows and even just walking down the local high street, I had only encountered inquisitive comments and encouragement.

Moving up a gear and progressing from being barefoot to running unshodden was another interesting and challenging step.

I still hadn't got it right. My calves still ached awfully after a run and the balls of my feet were prone to blister quickly if I wasn't careful and conscious of keeping my steps light. I realized there was far more to running barefoot than simply ditching the trainers. I should have done a bit more research. But then again, too much research and I may not even have made it to the starting line.

Didn't someone once mention that ignorance is bliss?

Some days had been much easier than others which I put down to a flux in emotional and hormonal states. I have frequently been called a sensitive soul; it transpired I also have sensitive soles. There had been many times when I had craved the safety and security of shoes just to alleviate painful insecurities and stop the ego nagging.

I had definitely become more focused. Walking barefoot organically evolved into a walking meditation as self-preservation ensures I remained constantly aware of my next step, over-riding constant inane chatter in my head. I had become more peaceful. The internal and external merged into a tranquil acceptance of all that is. It is what it is.

I had stopped eating meat and most dairy products. My body decided it didn't want to eat flesh anymore. I had not become fanatical about food, just aware; choosing to make my decisions consciously. Before eating anything I quietly asked myself if the food in front of me was going to make me feel better. It was simply a healthy, responsible communication between my mind and my body instead of allowing my mind its usual rampage and pillage.

It had been a small shift in consciousness that was paying the largest dividends.

I felt so much better, so much cleaner inside. I hadn't had any sinus problems since reducing the dairy products. I also consumed far less bread and felt far less bloated. If you want to see what bread does to your stomach, just put a slice or two in a little water and leave for a while. The congealed mass that you will return to is the same sticky mass that sits in your stomach making you feel lethargic and bloated.

It's not rocket science. It's taking responsibility for your finest asset. Your body is the only thing that you will take with you through the whole of your life; it makes sense to take care of it.

Now I ate to please my body, not my head or my eyes. Those are not always helpful allies when it comes to making healthy decisions.

I ate for energy, for fuel. The better quality the fuel, the more efficient the running of the machine.

I meditated more. I talked less. I listened more. I was more aware of the silences in between the noise. I no longer dreaded those silences or felt obligated to fill them. I felt more assured and authentic; more congruent. More comfortable in my own skin.

I am what I am.

As well as a few pounds lighter, I was a few friends lighter. There had been those who had not been that enamored by such big changes and disruption in their own familiar routines. It had been an organic pruning experience. I hadn't missed those who had chosen to leave.

And of course there had been new friendships made. I had been incredibly humbled by people's support and kindness on all levels.

It seems that people recognise and respect authentic endeavors. A reconnection with our natural self resonates with others.

I had met the Mayor of Funchal (now the President of Madeira), Miguel Albuquerque, at a Chelsea Flower Show social event earlier in the year and as a result had not only been invited to visit Madeira as a guest but invited to stay, also as a guest, at the beautiful Pestana Chelsea Bridge Hotel prior to my barefoot run. It overlooked Battersea Park. I was greeted like an unshodden celebrity by the managers Lynnette and Arturo, who were delighted to be able to make my barefoot experience a bit more comfortable.

Miguel is also a keen gardener and grows hundreds of roses in his own garden on Madeira. We got on well. I cheerfully accepted his personal invitation to visit the island and wrote what he later described as 'his favourite Travel Article about the Garden Island.' It can still be read as a blog on my website, The Magic of Madeira. It is a beautiful part of the world and pleased my penchant for islands greatly. It was also another reminder that the smallest of steps can lead to the biggest adventures.

Those magical manifestations were happening more and more frequently. I enjoyed and embraced them hugely.

Ask and it is given.

The tannoy crackled back into life bringing my attention back to the barefoot run which was about to start. The throng of runners started to jostle to line up under the banner.

The race would start in five minutes.

Except it wasn't a race, it was just a run; a fun run, I reminded myself, although my brother had insisted 'fun run' was a contradiction in terms.

I also reminded myself that, ultimately, the only race we run is with ourselves.

As is the case with a lot of my runs, be they of the organised or solitary type, the barefoot 6K passed in a bit of a blur. It appears I have perfected the art of zoning out or running meditatively.

Trying to pick my way through the razor sharp beech masts and spiny conker cases, while attempting to avoid bumping into other frustrated runners, was tricky. Just yards into the run, runners resigned themselves to donning Vibrams, Luna sandals, Vivo barefoots and other 'barefoot footwear'. Halfway into the second mile, I was joining them.

The fallout from the mighty trees overhead forced us into submission and into shoes. I found it ironic that the very recipients of this charity run, Trees for Cities, were making it nigh on impossible to complete. The trees were calling the shots. Nature's prerogative.

Like everything, no matter how good or how bad, how easy or how difficult, it came to an end. I crossed the line with other despondent runners. We took off our footwear and headed toward the coffee stalls comparing disappointment, experiences and reasons to be there.

I met with Blue Peter gardener, Chris Collins, as previously arranged. As a Patron of Trees for Cities, Chris had started the race and expressed his sympathy at the inhospitable additions to the route.

"The paths were as clear as a bell yesterday," he grinned. "You just can't do deals with nature."

We had a cup of coffee and a catch up before gathering for the Guinness World Record attempt for the largest barefoot race. It was hoped that 500 participants would attend to complete the 100m grass circuit to regain the World Record.

They did and I was one of them.

I had no idea how my barefoot path would continue or indeed if it would. But I was grateful for the experiences and teachings it had brought thus far.

It was my intention to continue to explore more of life in an unshodden state and continue to promote the advantages of spending time barefoot but day-to-day commitments had also built up and needed attention.

We were already into autumn, with winter knocking on the door. I had completed my challenge. The journey had proven to be of far more benefit than the actual challenge itself but was it a path I could continue along without a planned destination? A goal. Could I keep my new-found passion alight or would being bare foot get kicked into touch?

13.
Friends for life

'I had found a place where it was OK to get angry, in fact the angrier I got, the fitter I got.'

As a teenager, it had quickly become apparent that the time I spent at high school were not actually going to be the best days of my life. They were not even in the running.

I hated school.

I hated the hour long bus journey. I hated the vast echoing halls. I hated the emotionally retarded, angry teachers venting the frustration they felt with their own lives on to classrooms of unsuspecting children. I hated the undeserved amount of power that they were given. I hated the other kids for their seemingly easy lives. I hated being bullied. But most of all I hated the deep, debilitating fear that all of the above evoked.

So I didn't go very often.

Of course then when I did go, the experience was worsened by my very conspicuous presence. And that evoked unwanted attention and snide comments from the teachers as well as the bullies.

One girl was different. She would always be pleased to see me. She was also tagged as a difficult pupil. She was good company. She was Sally.

Sally was belligerent, bold and outspoken. She was also quite a loner. She was a confrontational loner; her almost aggressive behaviour made her difficult to get near for most people. She challenged people as a defence mechanism.

I was a retreating loner, also difficult to get near but because I chose to keep a safe distance from other people. My defence mechanism was avoidance.

We formed a good friendship based on our similarities and despite our differences. She was outspoken and I didn't speak much at all. But we both shared and almost craved a love of daring greatly and pushing boundaries. Our motivation was probably very different but we sought the same outcome - an escape from the day to day dross and difficulties.

Sally was sporty and competitive. I was always active but not competitive. We started going to a gym. Not only were we the only girls there but at fifteen we were the youngest.

It was unconventional enough for us both to enjoy and commit to. When our contemporaries were going out chasing boys, shopping and drinking, we went to the gym. I loved that time. It was only a basic set up in a small room as part of a Youth Club but I had eventually found an outlet for my frustration and anger. Two emotions that could be diluted by spending time with horses, but not vented around them.

I had found a place where it was OK to get angry. In fact the angrier I got, the fitter I got. It felt good. I had my allies; nature, animals and exercise.

Over thirty years later they are still my best friends.

My love of horses and a strong desire to avoid being bullied meant that I went to school less and less. I didn't sit a single mock O level and my parents had to attend a special meeting at the school to ask permission for me to sit my proper O levels. Dad had insisted. The teachers weren't keen. I was referred to as a lost cause.

Fuelled by the predictions that I would never amount to anything and the accusations that I was wasting people's time, I bagged eleven O levels with good grades.

Soon after the results arrived in the post so did an invitation to meet with the teachers to discuss my A level options. I went. Bemused with all the attention and the promise of a great academic life after all that had happened, I stood up and walked out of the room.

A little ironically, just when I was on the point of being accepted into mainstream society, I had managed to spectacularly ensure my continued isolation.

I have never been asked for my O levels, or indeed any qualifications, since.

I had already been earning a pleasurable amount of money by helping out at the racing yard in my early teens and upon leaving school at sixteen, I found earning money pleasantly uncomplicated.

You just had to turn up when you said you would, work hard and be polite. And know when to keep quiet. I had a degree in that.

Unfortunately, as I was sitting my O levels, another adult-induced drama had hit the racing stables making news nationwide. A horse from the stables had been raced as a ringer; that is when an experienced racehorse is switched with a novice. Unsurprisingly the experienced horse romped home by several lengths, a steward's enquiry was held, followed by an investigation and the scam was uncovered.

I had been blissfully unaware as I sat in the exam rooms and the whole episode had rather serendipitously flowed past me. The only fall out that affected me was that the stables closed.

I lost my sanctuary and my sense of belonging.

But I still had the work ethic.

My brother and I had been brought up to keep busy, which is one of the reasons I had got on so well at the yard. No task was too menial and no job too dirty for me to do and do cheerfully.

Dad used to tell us that if we had nothing to do, we could dig a hole in the ground and fill it back in. He wasn't joking.

We also enjoyed getting involved in the many labouring jobs around the house as Dad carried out various home improvements. By the time we attended high school, we could mix cement, lay and point patios, build and rake out stone walls and, of course, wash out a cement mixer.

My brother was completely car mad. While still in primary school he would go and help a local mechanic friend fix farm machinery. He could drive before he hit double figures and would often drive old cars around the field behind our house. He taught me when I was about twelve years old.

We may not have won any Emotionally Intelligent Awards at that age but we sure as hell had a mean arsenal of practical skills. And an ingrained ability to work hard.

The older I got, the more I thrived on hard work and the exhaustion it brought. Exertion = exhaustion = peace. It may not have been quite in Einstein's league but it worked for me.

14.
Allbutt Shoes

'Tom grinned, "We'll just have to make sure they do know you."'

It was the middle of November 2013; two years since I had run my extremely uncomfortable and disappointing barefoot 6K in Battersea Park and just three weeks before I was proposing to recommence my barefoot journey.

"I am going to try to be the first person to run across Wales - bare foot." I repeated my intention to Tom Wood, the Director of RWA.

"Training will officially start on the 1st of December," I continued.

"And I'll do the run over the weekend of the 31st May and 1st June. That will give me exactly seven months to train."

I wasn't usually that neat in my thinking but as I was trying to pitch something it seemed a good idea. I was hoping his company would be able offer some support.

Training could have just as easily been six months and fourteen days or seven months and three days as far as I was concerned but didn't sound quite as sharp or business-like.

I was also fully aware that seven months is right at the edge of my attention span. Anything over that and I would risk losing interest rapidly and would have almost certainly been lured astray by something far more current and interesting. I've never been good at planning too far ahead.

Seven months sounded good. Not too short as to be unbelievable and not long enough to be dismissed as not current enough. It also sounded like the amount of time it would take to train for a 45 mile barefoot run across a country.

Tom thought so too and was immediately supportive. "OK. It sounds good. I like the idea. Let me run it past the team and I'll let you know. Ha – run it past – get it?" He grinned. It was to be the first pun of many.

I met Tom in 2012 when he came along to my book signing of Allbutt's Almanac, a 'Gardener's Guide and Nature Lover's Companion'. We chatted about bees, the benefits of fresh air and health and well-being in general.

A few months later, as a result of pursuing my desire to deliver health and wellbeing workshops to corporates, we were meeting in his office in Blaenavon.

I knew I was fortunate to have unlimited access to fresh air and nature and to have accessed an authentic lifestyle.

I had also developed an innate interest in health and well-being and was passionate about personal responsibility and accountability, especially when it came to health - physical and mental health.

I had discovered that good physical health prepares a pretty good path towards a healthy mind and vice versa.

I had no medical or scientific training but I was enjoying the benefits of what I had explored, discovered and put into practice. When something works really well, I get very excited and want to share it by shouting it from the roof tops.

Talking about it in a conference room would be a start.

Tom was looking for an innovative twist to make their rather solid training days more palatable. A health and wellbeing presentation given in the 'carb crash' after lunch, by a rather exuberant, energetic gardener could possibly fit the bill.

It's all about timing.

When Alexander Graham Bell was looking for the telephone, the telephone was also looking for him.

Tom put his trust in me and the seminars were well received.

I incorporated a short session on the benefits of spending time barefoot, sharing what I had learned whilst training for the barefoot 6K a couple of years earlier.

It felt good to revisit and share some of the benefits and experiences I had picked up from my time barefoot.

Despite the benefits I had discovered and embraced, the more mundane aspects of 'life' had slowly over-taken in the same way as the sea creeps stealthily up the beach enveloping the sandy shores.

The focus had retracted from my feet and reverted to my head as various professional projects and personal paraphernalia took priority and my feet had once again, been rendered deaf and dumb in the safety of shoes.

As promised, Tom did let me know. It was good news.

RWA was happy to support me in my attempt to become the first person to run the width of Wales barefoot.

They would manage the marketing, social media and what I called the 'e-aspects', as well as generally chivvying me along as and when necessary. Tom works on the premise that these real-life projects are an effective way for his staff to learn new skills in fun and engaging ways as opposed to sending them away on conventional training courses. It also fitted in nicely with the development of their e-learning platforms.

It's all about timing.

So I had back up, a team - someone, other than myself, to be accountable to. For someone who was used to choosing the path of 'least assistance', it was a big step in itself. Our first meeting was to discuss the title of the project and the first steps.

The obvious choice for me was Allbutt Shoes. There was immediate hesitation amongst those with more experience of social media than I have - which was everyone else in the room. '#Allbuttshoes' could be mis-read as a hashtag on social media. All-butts-hoes weren't what we wanted to promote.

I was more concerned that people who didn't know me would miss the Allbutt/ all but inference.

Tom grinned. "We'll just have to make sure they do know you."

Allbutt Shoes was up and running.

A slick website was created and included several rather sobering facts of just what the 45 miles of Wales's waistline equated to:

The distance from Dover to Calais, 7,623 buses parked end to end and the one that made me tingle with anticipation... travelling to the edge of space.

Social media pages were born, pasted, posted, created, launched or whatever happens to them.

My barefoot challenge actually existed.

It had a presence on the T'interweb.

It was suddenly very real.

And ever so slightly surreal.

15.
A shared scarf

'We both knew we didn't really fit into conventional life as such, but it was ok. Or at least I thought it was.'

Leaving school at sixteen years old and with the stables closed, I indulged in a number of jobs locally and concurrently. I would cheerfully go from one job straight to another, often working fifteen or sixteen hours a day. There was no shortage of work and as I didn't enjoy socialising, I was free to work all the hours I wanted to.

I wouldn't wait for a vacancy to arise; I would simply decide what I wanted to do and where I wanted to work and then go there and ask for a job. If there wasn't one, I would keep calling and visiting until there was.

One would usually manifest quite quickly.

While other teenagers were waitressing in local bars and B&Bs, I had set my sights on the local Gourmet Silver Service Restaurant, Glan-Y-Dwr. The only one of its kind for miles around and which was renowned for only employing mature people. I called and knocked on their door over ten times before they agreed to give me a trial. I learned far more than silver service waitressing skills; it was posh and I loved it. Both the staff and the diners were well mannered, polite, and appreciative and I had found somewhere else I belonged.

Sadly the owners divorced, the business closed and another era of belonging drew to an end.

I then decided I wanted to work on the little community magazine that was produced locally. So I asked and was eventually taken on; I loved that too. Suddenly I had my own little column and a voice. It was a big step for someone who was used to keeping quiet.

Working hard had numerous benefits; there were obviously the financial rewards and the excuse that I was too busy to socialise or get involved in relationships and also I could avoid getting embroiled in the chaos that still ensued at home as Mum's bi-polar remained largely unpredictable. She took lithium and phenelzine as management medication but they brought problems of their own.

Sally and I still knocked about a bit, sharing a love of motorbikes, rock concerts and the gym. We did a parachute jump for charity and pledged to run the London Marathon one day. We would also sit for hours outdoors chatting about all sorts.

When the weather got cold, we shared a ridiculously long Dr Who-style scarf of hers that had also gone through school with us. It was so long we could both wrap it around our necks and sit or walk comfortably together. We even shared it under our helmets as I rode pillion on her bike. It became an infamous part of us.

She smoked, I didn't. I had a few boyfriends, she didn't. I worked, she didn't. We would argue about topics we totally disagreed on, usually falling about laughing at the ridiculousness of it all.

We actually didn't agree on many things at all. We didn't need to. There was a comfortable energy and stillness that existed between us; that cherished sense of belonging. We both knew we didn't really fit into conventional life as such, but it was OK. Or at least I thought it was.

On Boxing Day, when we were just eighteen years old, Sally committed suicide.

She hanged herself with our Dr Who scarf.

16.
Barefoot forward

'In fact I feel incredibly vulnerable,
very self conscious and a little bit silly.'

Clearing my first commitment hurdle by inches, I started training for my epic barefoot challenge on the 1st of December 2013 as planned. My account of it was written up, emailed to the RWA office and went something like this:

As the barefoot run will take place on the weekend of the 31st May and 1st June 2014, it makes sense to me to start the training today. In theory I now have seven months to train. I am not going to let the fact that three (maybe more) of those months are harsh, inhospitable, winter months, deter me in anyway.

I have checked out several local roads which may be suitable to start with - roads that have a good, gravel-free surface and that are also relatively traffic-free. They are of a premium. I have decided on the Llangattock mountain road which is also comfortably and comfortingly local.

My trusty cameraman and coach (both terms used lightly as he has only ever run a bath and is armed only with the camera on his iPhone), David, has used the tripometer on his car to gauge distances of 0.5 miles, 1 mile and 1.5 miles on the premise that if I find those distances comfortable I can retrace my steps and get 2, 2.5 or even three miles under my belt.

I am apprehensive.

This is my first barefoot run for three years and my first run for over six months; I have no idea whether I will manage thirty yards let alone three miles. It's cold; dry but cold. I abandon my shoes and socks in the foot well of David's car, and step outside immediately struck by how hard and hostile the tarmac feels. I don't feel very welcome in my naked feet. In fact I feel incredibly vulnerable, very self conscious and a little bit silly.

I take a deep breath and lope off; the intrepid cameraman and coach overtakes in his car to assume a suitable position to film me as I run past. As he disappears over the horizon I pray he hasn't gone too far.

That could be awkward.

My feet are already painfully cold and I am wondering at what level of pain I should stop. My mind is also protesting loudly. Both ends of my body are in revolt and the bit in the middle is looking to join them. This is just a trial after all, so would it be OK to stop after 100 yards?
I keep wondering and keep running.

I alternate between running on the grass verge and the tarmac, the grass is softer but colder. I choose the tarmac.

My feet feel like solid blocks on the end of my legs. I can't feel my toes. I try every visualisation I can drag up from my memory, from strolling leisurely along a hot, sandy beach to slipping my feet into warm sheepskin slippers but my feet are numb and not listening.

At least I can see the cameraman and coach in the distance; my pride will get me that far. As I lope past, a sense of achievement kicks in and I keep running. I realise that my feet no longer hurt and the tarmac actually feels pretty good. That also means I have the feeling back in my feet.

I think I am almost enjoying it.

I run and almost enjoy it for a mile and a half. I even consider turning around and retracing my steps to indulge my ego but common sense prevails and I decide to quit while I'm smiling. My feet are tingling and so is the rest of me.

One and a half miles.

On tarmac.

On top of a mountain. In December.

With bare feet.

Wow.

Even my coach is beaming.

I feel exhilarated and proud.

Suddenly it all seems very real and even slightly possible.

What I find particularly interesting about that first blog is my lack of ownership of the challenge; I referred to it as the barefoot challenge, not my barefoot challenge. It may seem a small point but it is indicative of the detachment I felt from the whole concept at that early stage. I felt like an onlooker, an observer. While I found it a little perturbing at the time, I was to learn that there is such a thing as healthy detachment. Remaining detached from an outcome is a positive attribute.

I dusted off my Vibram Five Fingers.

My initial and tentative steps into barefoot running three years ago had been completely and conveniently forgotten, until the pain in my calves reminded me. I start by running along the canal bank, as I am running after work in the dusk. It is the safest place to do so with an even terrain and absence of traffic.

I have to increase my fitness levels as well as toughen up my feet, so I don't feel any guilt at donning my Vibrams at this stage.

I wonder if running on the flat is annoying my calves.

My 45 mile run across Wales is going to be a far cry from a flat 6K in Battersea Park. This time, preparation is going to include copious amounts of research as well as time spent running.

But there has to be a bit of fun injected into it all too.

A friend suggests I borrow her Morris dancing ankle bells to wear whilst I run. As Christmas is just weeks away, it seems obvious (to me) to add a Santa suit. And that is how a couple of mountain training runs are undertaken - as a barefoot Santa with bells on.

It takes my mind off my feet.

My antics are captured on film by local entrepreneur and innovative cameraman, Marc Jones, and consequently find their way to YouTube where they can still be enjoyed!

Temperatures are low but morale is high.

I am enjoying it all; the runs are exhilarating and I am soon up to three miles comfortably.

The soles of my feet are already hardening up and I am amazed by how muscular my feet are becoming - it's not something I ever thought about but obviously I am developing the muscles in my feet.

They feel like monkey feet.

Far from embarrassed or worried, I am encouraged and proud. Being outside the box, alone, has always been an attractive option to me.

17.
Stepping into the ring

'I find exhaustion gives me temporary relief from life.'

I adore the following quote by Theodore Roosevelt: "It is not the critic who counts; not the man who points out how the strong man stumbles, or where the doer of deeds could have done them better. The credit belongs to the man who is actually in the arena, whose face is marred by dust and sweat and blood; who strives valiantly; who errs, who comes short again and again, because there is no effort without error and shortcoming."

It reminds me to step into the ring. Try as I might, I cannot stay in the ring. I don't know if it's feasible for anyone. But I have to have time out. Time to process emotional issues, time to just observe, time to lick my wounds, time to wonder.

"So what makes a 47 year old woman want to run across a country - her country - in her bare feet?"

It was a frequently asked question by many (and even myself) usually shortened to, "Why?"

It is still not easy to give a succinct and satisfactory answer, which, of course, is the type of answer most people wanted.

I wanted to reply, "Are you really interested or are you just asking?" But it was difficult to deliver without sounding rude. I certainly didn't mean any offence in any way; it's just such a hard endeavor to quantify. If they were just asking, then the answer would probably be too profound and protracted for me to offer and for them to receive.

It reminded me of training for trainer-clad marathons; occasionally I would run twelve miles to work, work the day physically outdoors and then run the twelve miles home as a way of fitting in my training.

People would screw up their face, shake their head in puzzlement and ask, "Why?"

I soon realised that those who had to ask were unlikely to understand any offered explanation anyway. That may sound arrogant. It's not about arrogance, it's about awareness. There are those who completely understand the need to push one's self to test and hopefully reveal new-found abilities, strength and brief moments of self-satisfying relief. They don't ask, "Why?" They just grin and say, "Well done." Those who are asking "Why?" have not yet discovered that often painful path to personal discovery; or maybe they have seen the signposts but declined to follow them.

In my frustration, I resigned myself to answering tepid questions with equally tepid answers; I heard myself saying, "I want to see if I can." "It's a personal challenge." "No one has done it before." And while all of those statements were rich in truth, they were lean by way of an explanation.

The effusive and more eloquent explanations would have included, "I find that pushing myself to the absolute edge of my ability keeps my mind occupied. I am afraid to let my mind wander, in case it doesn't come back."

"I am a loner. I like being alone."

"I am used to running away. I like running away. Running away always leads to relief."

"I ache for recognition, reassurance that I am an OK person. Self-doubt cripples me."

"Fear is a constant stalker and I am afraid I won't ever amount to anything. I will never be 'enough' - clever enough, thin enough, successful enough,"

And possibly the most painful, "I find exhaustion gives me temporary relief from life."

That's what I didn't say.

I often wondered if I was ever going to be able to 'do life' properly. I just didn't feel very good at living somehow. Couldn't seem to get it right.

As a teenager, my casual relationship with nature had developed into a clingy dependency. I spent as much time as possible outdoors.

I would go and talk to a neighbour's pony, breaking my silence by telling her all my problems, wiping away tears with her mane. Sometimes I would jump on her back, just sitting on her as she grazed, no saddle, no bridle and no fear.

It was one of the few places fear wouldn't follow me. It had no place there.

Visiting Mum in hospital continued and became part of the routine, as did the apprehension and vigilance when she was at home. The stigma attached to mental illness was diabolically rife and debilitating for us all. A lack of understanding presented itself as a lack of compassion; so many people were scared of it, the unknown quantity.

It was scary. Especially for Mum. We were hostages of the illness but she was on the front line.

And there was no cease fire when she returned home. That would be a protracted, carefully managed and exhausting episode in itself. At first, she would come home just for an hour or so, just a visit, a visitor in her own home. And then for a day. Always accompanied by a CPN, a community psychiatric nurse. And then for a weekend. That was always tricky too. She would be like a stranger. She had to adapt. We had to adapt. But nothing was ever said. We all got on with our lives, just making space for Mum. I would be relieved to be freed from my practical chores but weighed down by the vigilance needed in case it all happened again.

And of course it did, cyclically.

I lurched through my teenage years, dreading time spent indoors and around people and the amount of energy it took.

Being on high alert all the time, waiting for the next chaotic episode. Knowing it was coming but not knowing when. Knowing it was waiting in the wings, just waiting for its cue. Misery and mayhem were never far away.

At every opportunity I chose the company of plants and animals and the comfort of the great outdoors.

18.
We grow when it rains

'Quite simply, your body will not let you heel strike without cushioning, it hurts too much.'

Training outdoors, in winter, in Wales obviously brought challenges of its own. With Challenge Day set for the end of May 2014, the lion's share of the training had to be done during the winter months.

An alternative option had been to use the summer for training and to run the challenge in late autumn but not only did I not want to risk running across Wales in wintery weather, neither did I want to commit to spending the summer training on top of a hectic physical work load.

My work as a landscape gardener means I work outdoors all day. There is not much indoor work for us. We quietly envy the painters and builders that we work alongside in the summer months, as they slide off to indoor jobs for the winter. I am in my 30th year of difficult and exhausting winters. Whilst I have devised various methods of getting through eight hour chunks of hostile weather to appease the bank manager and keep the wolf from the door, adding another hour to a tiring day to accommodate a barefoot run was grim.

The winter of 2013/2014 developed into the wettest Welsh winter on record. Being wet is bearable; being cold is bearable; being wet and cold is grim.

It was also dark by the time I finished work.

Running barefoot in the dark brings more challenges which don't have to be explained. My Vibrams helped overcome this particular obstacle but I did experiment with a few totally barefoot twilight runs which actually proved to be particularly exhilarating on every level.

They felt a little indulgent, as they were very risky. A risk I could ill afford at this stage. Neither work nor training would benefit from a self-imposed injury.

I was already having to retrain to run in a different style as running without the support of trainers means adopting a totally different biomechanical method. I hadn't managed to conquer this for the 6K in Battersea Park three years ago. It was a short enough distance to complete at an ill-prepared-for lope.

But now I also had to train to actually run again. I hadn't run at all, in any shape or form, for over six months. Admittedly my work and lifestyle keeps me relatively fit but running is a funny old thing. You would assume that having some level of fitness and already having a few marathons under my Lycra belt, I would retain that running ability. It's not the case. Fitness is far easier to lose than to gain. Not only did my legs and lungs protest at being re-employed at this level, several other muscles joined in the chorus as they were not used to being engaged at all. My body wasn't happy with me at all.

There is a lot of information out there regarding the difference, biomechanically, in running with trainers and without. In a nutshell, the barefoot runner, one running without the support and cushioning of conventional running shoes, will naturally land on the ball of their feet as opposed to their heel. It's known as fore-foot running and heel striking respectively. Quite simply, your body will not let you heel strike without cushioning - it hurts too much. I was embarrassed to discover this information so late into my challenge. I had made the classic school girl error of assuming that running barefoot simply meant running with your bare feet directly in contact with the ground.

My brief barefoot experience had indeed been too brief to be of any real value with regard to running. I had no idea or concept of the impact that a considerable amount of running barefoot would have on the rest of my body. I had only ever run just over three and a half miles bare foot. And run was an exaggeration. It had been a 'light jog', a lope, a dog-trot. And that distance didn't need any technique, just determination.

Whether it was psychological or not, as soon as I neared 4 miles, my back hurt and my calves screamed, my thighs sulked and the bottom of my feet ached. The soles of my feet ached - for goodness sake how can the bottom of your feet ache?

I hadn't envisaged so much of my training this time would mean trawling through the internet uncovering reason after reason why most runners wore trainers.

And also discovering the reasons some didn't.

It was exciting but also daunting; time after time, expert after expert explained why the transition from running in trainers to running barefoot would take at least twelve months for the body to adjust.

I didn't have twelve months, I had six.

Names like Barefoot Ted, Daniel Lieberman and even Zola Budd took up residence in my head. Staunch and successful bare footers. I learned more about barefoot shoes, essential for transitioning. So maybe I would have to stick with the Vibrams for a bit longer. That would be OK.

My life was overflowing with the three Rs: reading, researching and running.

And all the time, lurching from determination and enthusiasm to despondency and doubt and back. As Henry Ford said, "if you think you can or if you think you can't, you'll be right."

I didn't think about it all the time - only when I was awake and if there was a 'y' in the day. It was becoming all consuming.

My first relationship with a barefoot shoe had been with the Vibram Fivefingers. I had used them to train for the Battersea Park 6K. They appealed hugely to my sense of humour and love of being different, with their separate frog-like toes. Likened to gloves for feet, they were comfortable and very, very flat. They deliberately lacked the cushioning and support of conventional trainers but the minimal rubber sole did provide protection for my skin soles. The neoprene glove also added a little warmth on those sub-zero days.

I had realised, the painful way, that bare feet don't have built-in treads to offer grip on wet ground. It was funny but it was also frustrating.

I tried to keep it all light. Tried to retain a sense of humour.

Not only was I training and working and researching, I was also blogging. I would use my running time to create engaging and inspiring blogs in my head. Finding the humour in running barefoot through sleet was tricky. Nonetheless, I would have a whole blog wittily dictated to myself and saved in my head during my run. After a quick bath and thaw, I would then settle down in front of the computer, look at the screen and forget every bloody word. I would freeze outdoors, defrost, and freeze again in front of the computer. Oh the irony. Oh the humour.

"Include lots of photos," Tom had advised. "They make a blog interesting."

So I used Boxing Day to make the most of an impromptu photo shoot with a friend, an iPhone and a pair of boxing gloves I happen to own. A Boxing Day Run was the title. And of course I ran. After the photos, without the friend, without the iPhone and without the boxing gloves but with beautiful blue skies and sunshine. Sunshine on Boxing Day. Yippee. It was a great run; energy and enthusiasm were abundant. I passed four other runners. It felt good. We were all in good spirits - and they were all in trainers.

19.
Boxing clever

'If I had been labelled 'difficult' as a teenager, I was about to be considered to be 'nigh on impossible' as a young adult.'

Sally's death had completely floored me. We were just eighteen years old. We had survived school and a lot more besides. We had talked and walked through some very dark times but we had also walked out the other side of them. Or I thought we had.

I was astounded and angered on so many levels.

Angry with everyone who was unaffected as they just continued to get on with their lives. Didn't they know what had happened? My world had stopped turning - why hadn't theirs?

And angry with all of those people who had been affected – the ones who knew her personally and the ones who had been involved with her professionally. Why hadn't they stopped it?

I was angry with myself - why hadn't I seen it coming, seen the signs?

And even angry with her.

That's a lot of anger.

As the anger escalated so the guilt gnawed away at me too. So this was another side of that Black Dog: depression. I had grown up with Mum's depression. Ye Gods, I was familiar with depression so why hadn't I noticed Sally's suffering? What a failing. What a failure. I should have known better.

I lurched from feeling like I would explode in a big ball of angry flames to feeling like I would just dissolve in a sea of corroding tears.

If I had been labeled difficult as a teenager, I was about to be considered to be nigh on impossible as a young adult.

Riddled with grief, guilt and anger, I badgered a friend to let me train with him. Gary was a young boxer who would in fact go on to become a Welsh Middleweight Champion. But then he was unknown, just focused, obsessed and driven. He wanted to be the best he could be. He was fit, cool and confident. And I wanted to be like him.

He knew of Sally's passing and told me, "Suicide is a way of making sure life can't sack you. You just quit. Just remember you're not a quitter."

Gary was a little bemused by my determination and eventually agreed to let me train with him as long as I kept from under his feet.

I did. I could do invisible.

The gym was rustic. His father's concrete garage had been cleared to accommodate a few appropriate props. There were a couple of skipping ropes, free weights, scaffolding poles, car batteries and an old front door propped against the wall for sit ups. It was cold, basic and even had a slight air of hostility as Gary and his mates trained in a serious silence. But I felt that familiar bliss of belonging.

Gary would make me skip for twenty minutes as a warm up; when I asked what was next, I would be told, "do another twenty minutes." Before long I would skip for over an hour at a time. I soon built up to over 100 sit ups too. I was fit.

That dark, cold garage provided an unlikely sanctuary. I liked the company. There was a rawness and honesty about it all. And I appreciated the fact that I didn't have to contribute to any conversation.

It got me though a very dark period in my life.

It was my preference for solitude and distaste for people which led me into gardening as a profession. Both of my grandfathers and even great grandfathers had been professional gardeners and my father had trained professionally as a landscape gardener at Pershore.

Amid the melee of jobs I had after leaving school, my father insisted I got a sensible qualification, something I could always fall back on. I chose a secretarial course. It also made sense as I wanted my own business one day. I joined the course late and completed the exams early. I also chose to keep working and earning money whilst I studied. I fully embraced the learning but got bored quickly, always reverting to staring out the window longing to be out in the fresh air.

I soon had a clutch of secretarial qualifications to add to my redundant O-Levels: Teeline shorthand, typing and a host of other certificates that proved to be about as memorable as they have been useful.

I had them all but no desire to use any of them. I had to find something I wanted to do. My smorgasbord of jobs was all very well but had no real long-term attraction.

Dad and I sat down one Sunday morning with a pen and writing pad and went through my career possibilities methodically.

I had to be self-employed, I was tiring of having too many bosses and my favourite jobs being lost through no fault of my own. I had to be outdoors. My work would have to be physical, challenging, creative and preferably with not too many people involved. Working with horses was my preferred choice but there were too many people involved and my experience of most of them hadn't been favourable.

Gardening was the only thing which ticked all the boxes.

I have since come to believe that you don't choose gardening, gardening chooses you. It tapped me on the shoulder over 30 years ago and I have always been immensely proud to be on nature's payroll.

A horticultural course at college was soon ruled out. There would be too many people and I would be spending instead of earning. Instead I would use my already substantial savings to buy a van and a few tools. Dad would help out with technical advice and at least I could now type out my own flyers and invoices.

I had enough money saved to buy a little Minivan, a second hand Flymo, a couple of hand tools and to put up a card in the Post Office window.

I was on my way.

I have since shared this following quote many times at various business meetings I have been asked to speak at:

"I established the business simply by turning up on time, sweeping up and shutting the gate as I left."

It was that simple. It was apparent that I enjoyed what I did in-between too. Passion is a powerful skill to have.

I was soon reinvesting my earnings in a strimmer, a hedge cutter and most of the Dr Hessayson 'Expert' books. A lot of the gardens I worked in had shrubs and plants that we didn't have at home and with which I wasn't familiar. It was a stumbling block.

Because I didn't know what they were, I didn't know what to do with them.

Dad came up with a plan.

Every Sunday morning we would go to a client's garden and plan what I had to do there during the following week. We took little pieces of different coloured wool. Shrubs that had to be cut hard back had a piece of red wool tied to them, things that could be cut back just to shape had blue wool attached and yellow was for the condemned. Copious notes and sketches were made and adhered to.

Dad was well known and well respected locally and clients embraced this rather unconventional approach with kindness. It was another advantage of a small Welsh community. I adored my new job. People were pleased to see me but left me alone to do my work. The work was challenging, both mentally and physically. I was learning and earning. I was also far too occupied and far too tired to have to address the less successful aspects of my life. I still missed Sally dreadfully and every single day wondered if there was something I could have done to prevent her taking her own life.

I still spent as much time as I could training with Gary.

Mum was still in and out of hospital.

I found relationships nigh on impossible. I still battled with low self-esteem and still never felt that I was 'enough'. I wasn't short of attention from the opposite sex and could cope with that easily, but affection was different. That evoked all sorts of emotions I didn't feel comfortable with; insecurity, jealousy and ultimately a fear of being rejected were all too confusing and uncomfortable. At least I could blame my successful workload for failing romance. I was just too busy. Too tired.

Wanting to save money on advertising, I approached the local newspaper and offered to write a regular gardening column for them if they put my advert next to it. They agreed. My first column had to be impressive. It was springtime and I wrote about 'scarifying' the lawn - removing thatch and dead grass to improve the health of the grass. It sounded good. I used the word 'scarify' a lot. I was young and ambitious.

So was the typist at the newspaper. Unfortunately she wasn't a gardener and typed up my very first column substituting the word 'scarify' for 'sacrifice'.

My very first gardening column advised readers how to, and gave the benefits of, sacrificing their lawn.

It was nearly the end of a wonderful writing career that had only just begun.

20.
Down to earth

'I risked criticism, rejection and ridicule. '

There is nothing like running barefoot in the rain and wintery wind with nothing on your feet to get you noticed. People were starting to talk. Most were inquisitive, some just supportive and some rude.

It was a step too far for many, especially in Wales. Bare footing is not big here.

But it does have a much bigger following than I expected, which I discovered largely via social media. And not just bare footers but barefoot runners. It does seem that even bare footers can be split into two camps; those who run barefoot or in minimal shoes and those who choose a barefoot lifestyle, often without running at all. Obviously the two overlap a little but are generally quite different. I found that the barefoot runners are usually those who used to run in trainers but after struggling with various running related injuries, have chosen to run without the traditional support and cushioning of trainers.

People choosing barefoot lifestyles are a smaller group and can be a little dogmatic about their choice.

Not surprisingly, I found myself with a foot in each camp.

Thankfully, they are, on the whole, a friendly, encouraging lot. There are several social media groups that were incredibly helpful with practical advice for barefoot beginners as well as those a bit further along their path. I am not naturally a group person but found it reassuring and refreshing to have such a concerted injection of positive advice and guidance, often based on actual experiences.

I think it's fair to say people are more disturbed by bare feet than other areas of bare flesh. We are all used to being exposed to women's bare midriffs, men's bare torsos, bare legs, even builder's bums. But witnessing bare feet in an unconventional environment seemed to make a lot of people uncomfortable.

I believe it is fear based. The thought of stepping on something sharp or treading in dog mess is what most people think of when confronted with feet lacking the protection of footwear.

For me, one of the biggest hurdles was also vulnerability but not to sharp objects or dog muck; I just felt incredibly exposed. I risked criticism, rejection and ridicule. It took courage to dare to be different and when you take courage you automatically risk feeling vulnerable. American author and public speaker, Brene Brown, has spoken beautifully on this subject, the power of vulnerability. I found spending time barefoot reminded me a lot of teachings I had heard and I believe it is an organic method of enhancing our ability to learn and make changes. It is as though it interrupts our neuro-linguistic programming naturally which then allows for new concepts to be explored and embraced with greater ease and speed.

Expose your sole and you expose your soul - to yourself. It is like plugging straight into a universal supply of authenticity and congruency.

Having my bare soles directly in contact with the ground tremendously improved my connection with the earth, and I don't mean the dirt, I mean the earth itself.

I think it is that particular connection that is the one that we are all subconsciously drawn toward. It's natural.

There is a lot written about the need for and benefits of reconnecting with nature but I don't think we can ever lose that connection; it is about rediscovering it.

And spending time barefoot is the quickest way to do it. I keep reminding people, you don't have to be bare foot all the time; just spending ten or twenty minutes a day walking barefoot somewhere safe will yield results. You can even sit down with your bare feet on the ground – no walking necessary. It doesn't get any easier, why wouldn't you try it?

There is much written about the benefits of earthing or grounding.

The earth has a constant negative electrical potential on its surface. When we are in direct contact with the ground, barefoot for example, the earth's electrons are organically conducted to our body, returning it to the same electrical potential as the earth. This therefore makes us feel more aligned or connected. Because we are!

There are numerous documented health benefits including improved sleep, reduced inflammation, improved mood, less anxiety, reduced heart rate to name a few.

Due to our rubber or plastic soled shoes and sleeping up off the floor, we no longer have the natural electrical connection to the earth that our ancestors had - unless you take the time to do so.

I believe that we also receive feedback from the earth by doing this. Without the constraints of shoes and socks, our feet become a sensory organ.

Because most people associate being barefoot with relaxation, whether it is at home or on holiday, it is easy to claim that explains any improvement in mood and well-being. But what if our feet are acting like a plug and the ground like a socket? The energy would flow like electricity.

Put shoes back on and you've dropped that connection. Put your feet in shoes and socks and they become blind and deaf, devoid of powerful external sensations.

I was at the rather unexpected and unconventional stage where I put my boots on in order to disconnect.

For me, it demonstrates just how far from our authentic paths we have veered; most people take their shoes off to relax but as a result of spending so much time barefoot, I was having to put my boots on to zone out. Relaxing meant not having to be constantly attentive to my surroundings or the terrain beneath my feet and also not having to explore and address the constant messages or teachings that I was receiving from the ground.

While filming one day, a BBC director shared a lovely story. He told me, "I remember years ago, spending a summer with my family in a caravan on a Welsh coast; the kids were young and we had a fabulous time. We were all barefoot most of the time and now, with families of their own, they still refer to it fondly as The Summer Without Shoes."

The simplest way I can describe it is that it is an organic and authentic way of recharging your batteries. Another analogy that struck me was that of a plug not quite being in a socket properly; when it is only half in you get flickers of a connection which would equate to the occasional 'a-ha' moment or thought but when the plug is pushed in properly the connection is consistent, providing a steady stream of 'a-ha' revelations.

Another interesting discovery I made is that dirt contains lecithin, a mood lifting chemical which is also found in chocolate. It can be absorbed through your skin. Maybe that's why various polls suggest gardener's are amongst the happiest with their jobs. It would also contribute to that feel-good feeling you get from being barefoot in nature.

Whatever the reasons may be, whatever reasons you choose, whatever reasons resonate with you, I highly recommend you just try it.

Free your feet and you free your mind.

Despite making strides into my self-awareness, I was still tip toeing into this new world. The rest of my life was still hum-drumming to a more conventional beat - there was work in its many guises and bill paying and housekeeping and animal management but there was also a new door opening and I was walking through it - unshodden.

It was proving to be an exciting and enlightening chapter in my life.

21.
Knowing a man who can

'Mostly on a Sunday and
often very early on a Sunday morning
when there weren't many people about.'

Aged eighteen, I had lost my best friend and my faith in human beings in general. I was unable to sustain a relationship, unable to trust and unable to care too much about it.

However, I had my own business, my own van, my own column and I had my own home. Whereas there were obviously aspects of life that I couldn't deal with well, there were various areas in which I was comfortable. They were all areas in which I 'paddled my own canoe'. Luckily I enjoyed my own company. I would still fight with the notion that I should be more sociable and accepting of people but it didn't make me happy. I would revisit the option now and again, encounter all sorts of emotions that were diabolical and debilitating and retreat at a rate of knots. Spending time in company seemed to dissolve my authentic self. I became a person who I neither recognised nor liked. Each retreat would be a little further back as I had to recalibrate to my authentic self.

There was safety in solitude.

Hand in hand with the preference for solitude strolled the desire for independence, so I had bought a mobile home and got permission to site it in the field of a local farm.

My father may not have been the best father in the World but he was a bloody good friend and long before I became blissfully aware of the fact. He would absolutely flatly refuse to help with anything financially. If my brother or I wanted something we had to work and save for it but he would pull out all the stops to help in other, more practical ways. My brother and I had to pay for driving lessons, buy our first car and pay for the tax and insurance. However, Dad drove us to the nearest town so that our lessons would be spent in the town itself instead of being wasted on a pick up trip and he drove us around to look at various cars before we bought one. It was only much later in life that I appreciated the value of this approach.

And he knew everyone! If he couldn't do, find or fix, he knew someone who could.

I had bought the mobile home from Hereford and had to get it 40 miles back to Crickhowell. He made a few calls and soon had a posse of mates help me move my new home. On a Sunday. He could make anything happen. Mostly on a Sunday and often very early on a Sunday morning when there weren't many people about.

A few more favours were asked from a few more mates and I soon had electricity and water to my new home. I created a little garden and bought a puppy. I had my own home. My own life. My own business. Even my own dog. I was independent. Safe.

I was also incredibly, achingly, tearfully, wretchedly, mind-numbingly tired.

22.
Run and become

'I bloody hate running.'

I am actually not terribly keen on running. Never have been and don't ever really expect to be. Nor do I find staying motivated easy, just necessary.

But I am hooked on the feelings that running (and being motivated) provide. On a good day those fabulous feelings can be enjoyed whilst you're actually running; on not such a good day they can put in an appearance after the run.

Without exception, on every run I take, the first half an hour or couple of miles consists of miserable, negative chatter and moaning. "I don't want to do this. Why am I doing this? Why am I bothering? I hate this. It's too cold/hot/wet/late/early. I'll turn around at the next bridge/gate/ junction."

Or just, "I bloody hate running."

At the start of a run, my legs feel like they belong to someone else and I run 'lumpy'. Then eventually, and thankfully, I find my stride. I sink into it and start to leave the negative chat behind as the endorphins take over. It feels great. I start to glide. It's smooth and relatively effortless, everything starts to work in harmony and I feel I can run forever. It's that 'Forrest Gump' feeling that all runners will recognise.

When I first began running seriously, I began in trainers. Sally and I had always said we would run the London Marathon together one day. Now she had gone, I felt an obligation to do it in her memory.

It took me fifteen years.

It took me fifteen years to even start running.

A year after I started, I ran the London Marathon. The only reason it took me so long to honour that commitment was because I hate running. I had told myself I wasn't a runner; I could walk for thirty miles but I wasn't a runner. I couldn't run. I wasn't built to run.

I can talk a good talk and had pretty much convinced myself over the fifteen years.

And then something shifted. I wanted to run. I didn't have to like it but I wanted to do it and I wanted to fulfil my self-imposed and embarrassingly postponed obligation to Sally.

I started to run along the canal. It was local, quiet and flat. In theory I could run from one bridge to another and then walk to the next bridge; run a bridge and walk a bridge. It would be an easy and gratifying way to monitor my progress. An easy way to start. In theory.

In practise it wasn't. I couldn't even run the distance between one bridge and another. My lungs burned, my legs had turned to lead and I was unhelpfully disgusted with my efforts. However, even over my banging heartbeat, I could hear the slap of the gauntlet being thrown down. The challenge had begun.

I would learn to run.

I was on my way to being a runner, like it or not.

My theory worked even if my legs didn't. Starting with a mix of walking, jogging and running worked well and I soon noticed that my distances became greater and my pain diminished.

As with so many things I do in life, it was only after I congratulated myself on finding something that worked that I found out it also works, and has worked, for millions of other people too. It's called Fartlek Training. Known rather affectionately as Fartlek, it was developed in the 1930s, comes from the Swedish for 'Speed Play' and combines continuous and interval training. Fartlek allows the athlete to run at varying intensity levels over distances of their choice. This type of training stresses both the aerobic and anaerobic energy pathways. As well as being effective with regard to improving my aerobic capacity, I found it also helped to keep my mind occupied, so it spent less time asking, "Can we go home yet?"

Puffing and panting I would lope between one bridge and another, then walk briskly to the next.

"Just one more bridge," I would promise myself, until I really had reached my limit. And then of course I had to turn around and do the same distance to get home.

I was a little annoyed to realise that my brisk walk was actually the same speed as my slow jog but my body only seemed to object to the latter. But slowly, mind numbingly slowly, the distance I ran became greater than the distance I walked. I felt fitter, stronger and happier than I had for some time.

Running, just me and my dog, gave me permission to be alone and also to disconnect from other people without seeming too freakish or aloof. I had an excuse to be alone. I began to look forward more and more to this me-time and chose increasingly remote routes.

As my fitness improved, so did the scenery. You can't go far in Wales without having to encounter a mountain and therefore hill running soon became necessary. It actually proved more enjoyable than the monotonous flatness of the canal and I was also far less likely to see anyone else.

I still loathed the first half an hour but began to obsess about the buzz that inevitably followed, however long or short it was. It was great. I felt I belonged. Belonged to my own body, to myself.

I eventually ran the London Marathon in 2004. As I passed the Cutty Sark I was overtaken by two huge purple rhinoceroses and around the 25th mile by Fauja Singh. He was 93 years old. In the previous year, this amazing British Indian centenarian marathon runner had set the marathon record of 5 hours 40 minutes for his age group.

For my first marathon, I had erred on the side of caution. As I jogged past runners sobbing on the side of the route unable to finish for one reason or another, I prayed my body and mind would make it to the end. There was no need for speed, just endurance. I ran through the finish just behind Fauja. I had been slow but I had finished. That's what mattered. It was a major milestone.

I was, of course, still wearing trainers at this time. Little did I know that one day I would be the first person to run across Wales without them.

23.
Staying one step ahead

' I believe that depression
starts with suppression.'

I have always worried about getting bi-polar. Mum was in her thirties when she was diagnosed. It seems it can strike, develop, manifest or maybe just become apparent at any time.

During my teens, I researched the illness endlessly, using the library in the pre-internet days. I would even make a list of questions and then make an appointment with a GP to discuss them all. I would have been about fifteen years old. My own emotional snags made me nervous. Did I have it? I would keep questioning, keep checking in with myself and researching.

"It is thought bipolar disorder may be linked to genetics, as the condition seems to run in families, although it often appears to skip a generation.

Studies report rates of bipolar disorder between 4% and 15% in children with one bipolar parent, compared to 0% to 2% in the offspring of parents who don't have the disorder.

However, no single gene is responsible for bipolar disorder. It is a complex condition with multiple contributing factors and there is also growing evidence that environment and lifestyle issues have an effect on the disorder's severity."

That sort of research filled me with fear but also resolutions; I would not have children of my own (a decision my brother also made independently) and I would keep one step ahead. Keep on the move. If it couldn't tag me then I couldn't be 'it'. A sort of emotional 'off-ground-touch'.

As a result I would constantly monitor my mind; keep it occupied and engaged. I was always analysing my behaviour, my thoughts and my intentions and, of course, those of others. Busy judging others, I was terrified of being judged myself. I became very guarded emotionally; it was safer that way. Or I thought it was.

As time has gone on I have become more rational. I have read and researched extensively about personal growth, self-management and the spiritual aspects of self-mastery. A little ironically, I now have my very own little library at home consisting of a considerable amount of 'shelf help' books.

The fear of developing a mental illness still prowls around my mind occasionally but it is dormant more often than awake. When it does stretch, yawn and turn up, rather than try to run away from it, I let it walk alongside me checking in with it now and then. It is no longer a threat, it is a teacher. I have learned so much from it and continue to do so. I respect it but will not relent to it. We have an understanding. All the time I am congruent, mindful and act with integrity, my path avoids the deep, dark gorges of despair. Should I become mean, thoughtless or careless, then my path begins to crumble a little and my way forward becomes harder to negotiate. It is a reminder to return to the congruent path.

Through my own experiences - my experience with Mum and, to a lesser degree, Sally - I believe that depression starts with suppression. Thoughts, feelings and needs which deserve to be expressed and addressed are suppressed for various reasons. They are subdued and quelled.

If left unattended, suppression deepens into repression. The very same thoughts, feelings and needs become more pronounced as they yearn to be recognised; they get harder to ignore.

Suppression is no longer strong enough, they need to be repressed. We cover them over with more and more materialistic rubbish and untruths, just to keep them hidden and to avoid dealing with them.

And if still left unaddressed, then repression evolves into depression. Depression is that deep dark rotting mess that results when you keep composting unresolved issues.

Sadly it seems that the symptoms of suppression and repression are often ignored; it easier to disregard them as inconvenient emotions. It is only the depressed state that is finally given attention. By that time, suppression and repression have ingrained the negative force into our subconscious and it is harder to address. The rot has set in.

Prescribed drugs become the 'answer', the solution. Of course they do not solve anything; they can't address and treat the cause, and they just mask the symptoms.

I get quite despondent to think that such suffering is born out of an inability to speak up for oneself, a lack of regard for one's views or the courage to express them and often through a lack of willingness from others to consider another's feelings and values.

I understand all too well how it can happen but get frustrated by the fact that it doesn't have to!

I have had no medical, scientific or formal training or teaching of any kind. This is purely based on my experience and subsequent interest in mental health and in particular the lack of regard for it by the majority of people.

Whilst I am achingly passionate about positive motivation and personal accountability, I am also intrigued by the triggers of apathy, self-sabotage and the methods of fear management.

Avoid the latter and you are well along the road to the former.

24.
Food for thought ... and energy

'Despite the hype,
I thought they were the Devil's food;
they made me feel quite ill.'

Tiredness has always been my ally and my enemy.

Thomas Edison once very wisely remarked, "The first requisite of success is the ability to apply your physical and mental energies to one problem without growing weary."

He was right. If I was going to run 45 miles in a weekend I would need to give myself the very best chance that I could. Amongst other things, I would need advice on nutrition.

My very first barefoot episode had lead to me cutting out meat and most dairy products. It had been difficult at first; I had to learn a new way of shopping as well as cooking. But slowly the new habits had become ingrained and I ate reasonably healthily. However, I felt it was more luck than judgment and when I was at my most debilitated I was sure valuable extra energy could be mustered from mealtimes.

I was introduced to Bristol-based nutritionist George Cooper by my Qi Gong teacher, Tony Dove.

I 'devoured' George's website, 'Be Your Own Nutritionist.' I bought his book (of the same title) and then arranged to meet the man himself.

He is a serious soul but I have come to appreciate that nutrition is a serious subject. Those who take nutrition seriously do tend to be serious people. I also appreciate that the word serious can also be interchanged with 'dedicated' and 'committed.' It baffles me how we get through life with such little knowledge and regard for the fuel we put into our bodies.

I, for one, am dedicated and committed to it. I have always expended a lot of energy; my work is physically demanding but so are all the things I do for recreation. In addition I am outdoors all the time so I also burn 'fuel' to keep warm. I use a lot of 'fuel.' I like to think that if I was an engine I would be a V8.

Years ago, my enthusiasm seemed to generate all the energy I needed but that is no longer the case. I respect the value of food and the necessity for the right type of food to best serve my body and my mind. I want to get the best performance out of myself so I am prepared to invest in the right type of 'fuel', sourcing it and preparing it. And if I know it is good for me and that I will enjoy the benefits then I will also enjoy eating it.

I have always found the hardest part is actually sourcing the right information. There is endless (and often controversial) advice available to those who are trying to lose weight but not a lot written about the benefits of food for optimum performance and energy.

I was very grateful to find George.

He chose a Chinese restaurant for our lunchtime meeting, the Dynasty in Bristol. I was slightly apprehensive but ridiculously excited to be having lunch with a nutritionist. What could be better? It was like learning to drive with The Stig.

Perusing the menu, George asked a few questions, "Are you hungry?"

"Always."

"Can you eat as much as a man?"

"Always."

"And other than meat, is there anything you don't eat?"

"I avoid dairy products when I can."

George ordered 7 small dishes and a bowl of Congee which was to be eaten at the end of the meal. He confidently explained that although I would feel completely full after the Dim Sum, the Congee would clear that full feeling as it speeds up digestion. After sharing the 7 dishes, we both realised I could easily have done them all justice on my own (plus another 7 probably) and I also enjoyed the Congee, which he had warned is "often an acquired taste." Like I said, if it's good for me, then I like it.

George explained that I was an 'ectomorph' body type (as opposed to an endomorph or mesomorph) and therefore I wasn't storing any fat reserves and burnt up energy quickly. Three good meals a day were going to be essential.

I explained that carbs often left me feeling bloated and sluggish, pasta in particular. And whereas most runners recommend bananas for energy snacks, despite the hype, I thought they were the devil's food. They made me feel quite ill. I'm sure they are a fine food if all you have to do is sit in a tree and scratch but not so fine if you are planning to run a few miles.

I explained that my biggest problem with nutrition was that I had been relying on trial and error to find what foods worked for me and the trial and error method can be long-winded (no pun intended). As I prepared for my 45 mile run I didn't have the luxury of trial and error time. I was also questioning my own intuition. I was getting frustrated and I didn't want to be wasting any more time.

George was very understanding and also validated my banana theory by telling me about an article he had written recently: 'Murray's Centre Court Diet is Bananas, says Nutritionist.' It turns out that bananas do take a lot of digesting in their raw state; they are not a beneficial food for a lot of people, unless cooked.

He also explained that fish soup, made with a stock of bones and heads of white fish, may be the most nutritious food on the planet. It has all the minerals and proteins from the bones and all the micronutrients from the sea and is a perfect tonic if your body is under stress. It's also great for generating energy. I had already tried the soup on his emailed recommendation and it had indeed been a complete contrast to the banana experience.

I had boiled the fish heads up, drained off the stock, added potato, parsnip, carrots and spinach, cooked and then blitzed with a blender. The result was rather thick, quite green and very fishy. It took considerable effort to eat the bowlful but trusting George, I did manage it. An hour later I felt as though I had eaten a bag of orange Smarties. I was bouncing! It was quite incredible.

I love the fact that food can do this. I love the fact that I know it can and even more, I love the fact that I am disciplined enough to do it! Result.

I felt exactly the same after lunch with George. I had literally re-fuelled; I felt great. It had consisted mainly of fish and steamed rice but all flavoured beautifully. I was going to be eating a lot of fish, rice (white) and green veggies. Almost every meal would contain chilies, ginger, spring onions and other spices, even scrambled eggs (the recipe is in George's book). Congee can be made in a slow cooker so that was also do-able and I could experiment enthusiastically with the fish soup.

George also reminded me that preparation and planning is the key to healthy eating. I needed to be thinking ahead about meals and often preparing them in advance. The slow cooker would be useful. I also needed to get into the habit of cooking in batches to provide meals for a few days or to freeze. My main problem was that I cooked enough for two or three meals but ate it all in one evening. My brother says I eat like a Drumming Bunny. What can I say? It's fuel and I'm more of a 4×4 than a Fiesta!

But I pledged to be more disciplined; I liked having a plan.

As snacks, George recommended salted cashew nuts, non-crystallised ginger chunks, buckwheat pancakes and nut butters. I later discovered nut butter on sliced apple works well too.

In his book he has a recipe for pancakes using buckwheat flour and almond milk rather than ordinary flour and milk which makes the pancake hard to digest. As he kept reminding me, it is all about the ease and speed of digestion; improving the motility of the gut.

In a surprising twist, he also suggested that I turn to vodka. Really! He went on to explain that I should add chopped ginger, cloves, cinnamon, cayenne pepper to the bottle and leave it to steep for two weeks before using it as a foot rub.

I would then be able to massage it into my feet before and/or after my run to stimulate circulation which would help recovery from the bitterly cold conditions. I was thrilled with the advice and very grateful; I had drastically underestimated just how cold the ground can be during the winter.

I was learning that life is all about being light. To run barefoot you must run lightly; a good mood comes from a light attitude and a healthy diet means eating lightly. And nothing weighs heavier than negativity, ill health and lethargy.

25.
Controversial and cruel

'As she gradually improved, so did her awareness of the horrific surroundings and the barbaric procedure.'

During the many years and cycles of sectioning and hospitalisation, the only thing that would eventually help Mum was electroconvulsive therapy. ECT involves sending an electric current through the brain to trigger an epileptic fit, with the aim of relieving severe depression and to treat mania or catatonia. The treatment is given under general anesthetic.

It was always used as a last resort, partly due to Mum's insistence against it. Inevitably, after months of ineffective medication she would become resigned to it. It is an extremely controversial treatment. There are many negative aspects but it transpired that ECT consistently had a positive effect on Mum's mood. The negative effects included large amounts of memory loss, although unfortunately not of the treatment itself. She still has some horrendous memories of the procedures. She recalls a Brigadier type gentleman administering the ECT as the doctors wouldn't. When he retired, the army took over his unenviable task in the continued absence of a willing GP. She remembers the bruises from the straps which restrained her during the convulsions.

As she gradually improved, so did her awareness of the horrific surroundings and the barbaric procedure. The fear associated with the latter episodes were to be etched more deeply in her memory, which is possibly why they were amongst the memories which avoided deletion by the treatment.

In general though, large parts of her short-term memory - and therefore the behaviour that resulted in her being sectioned - would be erased from her memory. It wasn't a bad thing; her manic phases caused an enormous amount of disruption and distress. She had enough to deal with without reliving those.

Not surprisingly, my parent's marriage was fraught and volatile. I recall my father absorbing an awful lot of abuse. It wasn't a pleasant environment a lot of the time.

During one manic phase, she threatened both my brother and I with a carving knife.

I recall her pushing me up against a kitchen cupboard hissing at me that she wished she had killed me at birth, when she had the chance. It was as though she was possessed.

Of course it was the illness but it was in the guise of my mother and as a young teenager I wasn't mature enough to separate the two. I took what I heard to heart.

And of course, when she was well, it was very different; she was kind, capable, compassionate and extremely creative.

Sadly, instead of enjoying these periods, I recall anxiously watching out for the signs of uncharacteristic behaviour and anticipating the next manic episode.

During one manic stage, Mum befriended a local fortune-teller. She became obsessed with the woman's predictions and forecasts. Mum came home one evening in a particularly agitated state. She told me and my brother 'in secret' that the fortune teller had told her that Dad was going to die in a car crash. She had described Dad's truck accurately and said it would be soon. I was thirteen and my brother was eleven years old. It was a heavy load to bear. The fortune teller adhered to this particular prophecy for years adding that it was difficult to interpret time in the 'other world'.

It never happened.

Obviously bi-polar played a significant hand in this cruel episode. It explained Mum's behaviour. I have failed dismally to comprehend the behaviour or motives of the fortune teller.

The same fortune teller had a grandson who was jailed for murder. To appease the fortune teller, Mum wrote to him regularly and through him was introduced to a fellow inmate, imprisoned for shooting his friend's mother. Mum started writing to him and began a 'pen-affair'. He was even granted a transfer to Cardiff Prison so she could visit him. I went with her on one occasion. It was on a par with visiting in Mid-Wales Hospital.

In order to keep it all from my father, she told me to pretend that the inmate was my friend. Dad was a postman at the time, so she couldn't hide the letters. The letters and visiting orders would come addressed to me with HMP stamped on the outside of the envelope; inside would be another sealed envelope for me to give to my mother.

My father was disgusted that I had befriended a prisoner and asked me to stop contacting him. Of course I couldn't and nor could I explain why. He shunned me for weeks. His disappointment and misplaced displeasure was heart-breaking. I was fourteen years old.

The whole harrowing episode eventually faded into oblivion as Mum was sectioned again. One of the bittersweet flavours of the procedure was that it organically enforced closure to many chaotic chapters. Mum was literally removed from the situation and indeed, society.

Not surprisingly, the relationship I had with my father, as a teenager, was not a comfortable one. Nor was it authentic.

Thankfully I did have the opportunity to get to know him properly later in life and as a human being, rather than a father.

It was a blessing.

26.
The crossing of paths

'Blaming social media for people's apathy or other traits is like blaming a fork for making you fat.'

As I continued training for my 'cross a country run' (a far cry from the school cross country runs which I loathed), my barefoot path was turning up some beautiful synchronicities and wonderful opportunities.

I am not a great lover of social media for various reasons, however I do believe it has many attributes if used sensibly and with integrity.

Blaming social media for people's apathy or other traits is like blaming a fork for making you fat. Social media has introduced me to some lovely people who I would have been unlikely to meet otherwise.

Take Glastonbury-based personal trainer, Michelle Blackmore, for example. We were introduced on Facebook by a mutual friend with whom I had previously run up Arthurs Seat with in Edinburgh, clad in light suits as part of a Speed of Light production.

Random on every level.

While she's not in my vicinity geographically, Michelle is certainly on my wavelength emotionally. We clicked after an email or two and she introduced me to one of my favourite 'energy secrets'. I already used chia seeds in cooking and salads but Michelle told me to add them to my water bottle when I ran. The consistency isn't great - it resembles frog spawn - but the results are worth it.

Michelle was also a huge fan of coconut water and its remarkable rehydration properties. I had already discovered this in my extensive research and always took it on my sporadic three day Woodland Fasts. When time and incentive allowed, I would wander off in to some remote woods with just my tent, my dog and my coconut water (with food for the dog obviously). No phone, no watch, no food and no expectations. I highly recommend it; the coconut water and the retreat.

I was beginning to feel a bit of a prima donna with a nutritionist and a personal trainer on hand even if they were both of the 'remote' kind.

And that is the Welsh-ness kicking in as it demands to know, "Who do you think you are with all these posh ideas?"

It was much harder for me to ask for and accept help and advice than it was to actually train. I have more experience of training.

As well as creating a running schedule and a weight lifting regime to help me build my core strength, Michelle was emphatic about the importance of rest days. She warned me there had to be rest days involved in training and they would be as essential as the exercise days. It is during recovery that the body builds and re-calibrates its improved muscle ability. It was a delicious relief to be given permission to rest.

Michelle's expert input made me realize the potential benefits of consulting a personal trainer closer to home. I was planning to run two marathons in two days. Across country, the Welsh countryside. There would be mountains and lots of them. I had access to reliable information about the 'fuel' or food I would need - surely I needed similar information about the physical aspects? I needed a personal trainer close to home.

27.
Not the answer

'I didn't even feel depressed, just hopeless,
hope-less – without hope.'

At just eighteen years old I was enjoying great success with my gardening work. A young female gardener was a novelty. We also lived in a small community and I was extremely well supported. Dad decided to leave the Post Office. He no longer needed the security that he had opted for in order to raise me and my brother. After eighteen years of doing the right thing, he could now finally follow his passion - landscaping.

We joined forces and became R J Allbutt and Daughter. I still don't know who was the most proud. We had the van sign written. A bigger van. A transit van. We bought more machinery, safe in the knowledge that my brother, who now was serving his apprenticeship as a mechanic, could keep it all running. We took on more work, more workmen and larger projects. I learnt the landscaping trade and garden design skills.

Garden design was not the in-vogue profession it is today. If we wanted to secure the landscaping work, we had to come up with a workable design. It was that simple. I loved it. All aspects of it, from the creative design side to the physical elements of operating heavy machinery. It was hard work but I kept up, mixing cement all day, laying slabs and digging trenches. I had a good sense of humour and loved working on site with the other workmen. I had a natural eye for design and a down to earth practical knowledge of what would work. I was soon giving design advice. Dad and I had a rather unpretentious motto with regard to design, 'Let the client tell you what they want and then tell them why they can't have it.'

I could also type and file. I was soon doing all the paperwork in the evenings. I was getting more and more tired and began misplacing my sense of humour. Bad weather made work hard. Being cold and wet was debilitating. I took on an evening bar job and a Saturday job at a florist, so I could cut down on working outside in the week. It didn't happen as planned. I kept up the same amount of hours outside as well as the additional hours indoors. I ached and struggled physically but at least I was still largely emotionally numb too.

At the end of January, aged just nineteen, after extra shifts at the pub, a busy Christmas at the florists, the anniversary of Sally's death, a particularly difficult period with Mum and still working every day outdoors in the harsh winter I discovered a boyfriend was cheating on me... again.

It was the final straw. I was exhausted. Too tired to even think straight. It all became insurmountable. Too difficult in that moment. All it took was a moment to make the decision.

I took an overdose.

It was a cry for help; I had been crying for help for an awfully long time. I had been labeled a difficult teenager, been blamed for causing my mother's illness, been bullied, lost my best friend, been disregarded and now discarded... again.

I didn't want to quit, Gary had been right, I wasn't a quitter but I had cried inside for too long and my cry for help became externalised. I couldn't suppress it any longer.

I didn't even feel depressed, just hopeless, hope-less - without hope. I was simply too tired to make excuses to people and for people.

In spite of being aware of the devastation and pain that Sally's suicide had caused, I was too tired to care.

A friend turned up unexpectedly and I was rushed to hospital.

When I came home nothing was mentioned about the incident and my father didn't speak to me at all for a week or so. Any pride he had felt had clearly dissolved. It was his way of dealing with it in hind-sight. He had enough on his plate.

Attempting suicide, no matter how half-hearted was a stupid thing to do.

I have learnt that nothing in life is ever that bad and no matter how bad something does appear to be, it will not last forever. Winter always passes to make way for spring and the obstacles of your past can become the gateways to new beginnings.

28.
Fit for purpose

'I hadn't even thought about the risk of injury and what that would mean.'

Despite being a big believer in the powers of visualization and a positive mental attitude, if I was going to run the width of a country - never mind barefoot - I would also need some more advice on the physical aspect of it all. So I popped to see a personal trainer I had used a couple of years previously. Mike James had been in the process of setting up his own business locally and I had been looking for ways to increase my energy levels to cope with a heavy physical work load. I had been warned that I was producing too much cortisol as a result of 'exercise addiction'. I had inadvertently created a vicious and debilitating circle. I had turned to Mike to help me strike a balance and had felt a lot better in a short amount of time.

So it made sense to turn to Mike again. He was now an advanced personal trainer with his own studio and had serendipitously recently qualified to teach barefoot running too. That was barefoot running with barefoot shoes rather than bare feet but nonetheless the principle and altered running style was exactly the same as the one I needed to master.

I had my first session with him at the end of January, just four months before my challenge and it proved to be the most sensible step of the journey I had taken so far.

I like Mike's style; he is confident, straight talking and effective. He has that enviable aura that one has when they have found their 'divine purpose.'

He is so very comfortable in his own skin.

He was also very busy, so we agreed not to use our limited time together to improve my running distance (I could do that alone) but to focus on style, flexibility and a few other things that would make me run more efficiently and safely. Most importantly, he would show me how to avoid injury.

Ye Gods. The cold hand of fear gripped me tightly; I hadn't even thought about the risk of injury and what that would mean.

Mike has a fantastic can-do attitude and a great positive mental attitude toward life in general. And that's an added bonus when preparing for extreme challenges.

I had run quite a hard two miler before my first session with Mike and my calves and quads hurt like hell as I arrived at his studio.

"They will," he cheerfully reassured me as I grizzled.

He put me straight onto the treadmill and recorded me running to gauge my style and inevitable errors. I felt ridiculously self-conscious and continued to moan accordingly.

"Just run as you would normally," he instructed.

I had my Vibrams on as instructed by Mike and was immediately conscious of how heavy and loud my footfall sounded. My Vibrams sounded more like steel toe capped boots. Mike confirmed this with a cheerful and almost triumphant, "Yep, you sound just like Nelly the Elephant."

He explained that it was the heel strike that was making the noise. I was running 'heel to toe', landing on my heel and rolling up on to my toe to push off. It is the most common but not the most efficient way to run. It is how we naturally run in trainers when we have the luxury of the support and cushioning they provide.

It was exactly what I didn't want to be doing barefoot, without the padding and support. It would hurt.

Putting his iPad (and evidence) to one side, Mike summed up my efforts: "OK, you look like a hunchback, sound like Nelly the Elephant and have absolutely no rhythm."

He added, "See how I made that into a compliment sandwich - two compliments with a criticism in between?"

"I didn't actually spot the compliments Mike, where were they again?"

"There weren't any. I lied". He shrugged. "We've got some serious work to do."

He told me to 'ditch the Vibrams' and then put me through a few exercises in front of those awful 'can't hide anywhere' gym mirrors; this was becoming far more painful than the ache in my quads and my calves. He quickly deduced that I'm reasonably flexible but have an extremely tight right planter fascia which would be very likely to cause an injury at some stage if not addressed. I also carry a lot of tension in my shoulders.

Now this really bothered me. It is a recurring theme in my Qi Gong classes. I keep getting told to relax my shoulders. And now Mike was saying the same. I didn't know how to. Damn. I really should know better. This was proving to be about so much more than my feet. Suddenly, and not for the first time, my whole body, mind and lifestyle was under scrutiny and being judged. Talk about feeling exposed and vulnerable and it extended far beyond my feet.

Exercises completed, mirrors rejected, I jumped back onto the treadmill without the Vibrams and Mike filmed me walking and running again.

I could feel and hear the difference immediately. I was running up on the ball of my foot, not quite tip toe but nearly. The amount of propulsion from the ball of my foot was also noticeable. I was lifting my feet rather than pushing off from them. Not unlike skipping. And I had been good at that once. It felt markedly different and Nelly was nowhere to be heard.

Mike explained the whole forefoot running principle and the fact that any stress would be transferred to the calves and quads instead of knees and hips. It was why it was considered by many to be a preferred method of running resulting in far less injuries. It triggered vague recollections from random research. Hearing it face to face was so much more effective.

It was indeed a major step forward and not before time considering that I had already completed one barefoot 6K and it was only four months before I was attempting to run 45 miles across Wales. I was encouraged and discouraged in equal proportions. How had I missed this essential part of barefoot running? How was I being so naive? I had not come into it via a conventional route. Most runners discover the benefits of running without shoes or in barefoot shoes as a result of sustaining injuries whilst running in trainers. Relevant research will lead you into the arena of forefoot running. I had sort of gate crashed the whole bare foot running party, fully expecting to run before I could walk.

I ramped up my research and re-read the infamous 'Born to Run'. I had read this whilst training for my very first trainer-clad- marathon over ten years ago and had even skipped through it again during my barefoot 6K training and had forgotten most of it.

I did however, immediately recall the surge of inspiration that it evoked. I wanted to run like a Taramuhara runner. I wanted that flight of foot, that lightness of stride, that amazing endurance, that buzz, that enlightened state, that expertise, that experience.

And I wanted it now.

The more I crammed my research the more disheartened I became. The general opinion of runners who had made the transition from heel striking to forefoot running was that it took in excess of 18 months to do so safely. I didn't have 18 months. I didn't even have 8 months.

Should I be postponing this whole thing?

29.
Blonde and beyond

'I can assure you that choosing a man for money is in fact the hardest way to earn it.'

In 1985 my father and I joined forces to become RJ Allbutt and Daughter. We were inundated with work. A father and daughter team was an unconventional concept let alone in the field of landscaping. Various articles featured in various newspapers and we enjoyed our local notoriety. Our personal relationship remained strained but on a professional level we were united by a mutual respect. On site we were good mates.

My brother had completed his apprenticeship and was now studying mechanical engineering at college in Swansea. He was as level headed and consistent as I was hot headed and inconsistent.

We were close though; brought closer by Mum's illness I am sure. She was still hospitalised on a semi-regular basis. The prescribed lithium would slowly build up in her system and turn toxic resulting in a 'high'. In this elevated mood she would decide she no longer need the lithium and would stop taking it altogether. A manic episode would ensue, with varying degrees of madness and mayhem, followed by sectioning, re-calibrating medication, debilitating depression and ECT before she would return home. The cycle was recurring approximately every two to three years and could take up to twelve months to correct. The uncertainty had become a certainty itself.

Being self-employed brought even more uncertainty.

I worried when we were busy and I worried that we might be quiet. I had never experienced a lack of work in my life, far from it. But it was still a possibility and my fear of never 'being enough' added fuel to those painful flames of self-doubt.

I gathered qualifications and skills like scatter cushions, working on the premise that the more skills I had then the more comfortable I would be, the less likely I would be to ever be out of work. They were all skills that would enable me to continue to be self-employed. Working for someone wasn't an option. That bridge had been burnt a long time ago.

The irony is, of course, that you actually have to rely on more people for your income and they all become your boss. I chose things that I could also diversify into during the harsher winter months should I need to.

I qualified in Reiki, Indian Head Massage and On-site chair Massage. The latter of which saw me demonstrating and giving treatments in the window of the Body Shop in Cardiff one winter. I also qualified in hairdressing. The tutors at Hereford College were incredibly flexible with the course content in order for me to attend and complete my training by attending one day a week. I kept the whole episode to myself. I have never felt so out of place as I did as a 'mature student' on a hairdressing course in Hereford College amongst a gaggle of preened sixteen and seventeen year olds. I qualified in twelve months.

My rationale was that if all else failed, at least I could cut hair on a beach somewhere. I also employed another killer tactic to disguise my lack of confidence. I went blonde. I figured that it would mean others would expect less of me. If that was the case, they could only be pleasantly surprised by what I delivered. I'm not sure that reasoning was ever proven but it certainly didn't appear to do me any harm.

I could do that myself.

Following one particularly bleak winter, I became incredibly ill. I had lost a lot of weight and working outdoors in hostile conditions had taken its toll on my health in general.

I was hospitalised and diagnosed with anorexia. I was just under seven stone.

And bloody furious.

I was not anorexic; I had absolutely no qualms about my weight or thoughts of being overweight. To the complete contrary, I simply couldn't put any weight on. I was exhausted and I simply could not consume enough calories to put weight on, let alone keep it. I already ate more than any man on site. I discharged myself and pledged to eat more.

It was yet another of my diabolical experiences of a highly unprofessional medical profession. During the whole frustrating fiasco, a GP advised that I take anti-depressants. He claimed that my weight loss was due to depression. When I refused, he said, "Your aversion to anti-depressants is an over-reaction to your mother's medical condition; you will be back for them one day."

He is still waiting.

But things had to change on another level. After chatting honestly with Dad we decided that I would get another job for the three or four worse winter months. It would be OK – I would be like a daffodil and come back out in the garden in the spring.

A friend worked for a Private Detective Agency based in Brecon. It was unconventional work with wages being based largely on commission and they were often looking for casual staff. That sounded perfect. I had no idea what working for a private detective entailed but was sure I could do it if it meant I had work indoors for three months.

I phoned and spoke to the boss. There were no jobs available but I was welcomed to call back some time to check if the situation had changed. I did. The next day.

He was somewhat taken aback but said that there were still no vacancies. By all means I could keep trying, as people did leave on short notice.

I called back the next day and every day for a week.

I started work the following week.

I was told that I had what it took to be a telephone tracer. I learnt all that I needed to fast. I loved the work but hated the lying which was a necessary part of gaining the required information regarding people who had defaulted on payments for goods.

I associated the lies I had been familiar with in my childhood with pain and undesirable outcomes. I wasn't happy lying. I did my three months and was offered a substantial incentive to stay. I had had enough of working indoors and of lying. I left, taking a lot of valuable lessons and experience with me.

Ironically, it also fuelled my distrust of people even further. Whereas having the knowledge was an advantage in some areas, the additional emotional baggage wasn't.

I was glad to get back outdoors and into a healthier environment, in every sense. My confidence had improved from the detective work and I was now pricing and securing jobs myself. One morning I went to the big house in the village to give a quote for maintaining the grounds. A millionaire was renting it. It was hot gossip in the village.

I got the job. He was an intimidating man. He quickly decided that I was the 'girl for the job' but was brusque and bullish in his manner. He told me that the van was not to be parked in front of his house under any circumstances and nor was I to have my break or a cup of tea on the premises. He didn't want it resembling a building site.

I wasn't bothered; he could be as weird as he liked. He was paying me.

That same evening I was working behind the bar in the local pub and served the brusque millionaire. He was very different, chatty, charismatic and complimentary about my work ethic. He apologised for being abrupt earlier, blaming a stressful business deal.

The very next day I was working in the local florist, my Saturday job. He walked in to order flowers for his mother's birthday. In true Hollywood- style he was absolutely charming and seemed somewhat endeared by me having yet another job.

"Do you ever have any time off?" he asked.

"No," I replied.

"Well if you do, I'd love to buy you dinner. Call me." He pressed his business card into my hand.

I didn't.

He became a regular in the pub on the nights which I worked and would often call into the florists on a Saturday; he even brought me a cup of coffee whilst I worked in his garden. He was kind, complimentary and charming. He reminded me to call him.

I did.

Within six months we were living in the Caribbean on the tiny island of Montserrat. He had completely swept me off my feet. It had been just like a fairy tale. Friends teased me about becoming a 'kept woman', having a 'rich man' to look after me.

I can assure you that choosing a man for money is in fact the hardest way to earn it.

After only a month in the Caribbean, as well as sporting a beautiful tan, I was nursing a swollen lip, two cracked ribs and a black eye.

And it hadn't begun there.

He had first hit me whilst we were still in my home village. Afterwards, he had been mortified, blaming stress and the pressures of big business deals. And, he added, I had provoked him. I had changed channels on the TV without asking him.

I am sure to this day, that if he hadn't inferred I had a part to play in his violence, I would have walked away there and then. It just wouldn't have made any sense to stay. But somehow, his reference to me being to blame for his inappropriate behaviour struck a chord. I had heard it so many times before amid chaotic circumstances: "This is your fault."

He must be right.

I must be wrong.

He bought me a Rolex watch by way of an apology and promised it wouldn't happen again.

He lied.

30.
Mind over matter; head over heels

'I had walked barefoot over red hot coals.'

It's hard not to think when you are running.

I occasionally run with an iPod but more often than not I have forgotten to charge it or more often just forgotten it. I prefer to run with my thoughts as company and endeavour to keep them motivational and beneficial.

Sometimes I will just run through plans and ideas for a client's garden or rehearse a speech I am due to give. But mostly I enjoy inventing little mantras and affirmations. While running barefoot, I would try to vividly imagine the conditions I wanted in an attempt to block out the conditions I had. When my toes had gone past numb and had ventured into the painful stage, I would imagine them in front of a warm fire or wrapped in sheepskin slippers. Simple and effective.

I have such respect for the power and potential of the mind.

There would be the odd incident that would put a spring in my step. One day, while running along the canal bank, I passed an adult with a small child. I overheard the child ask, "Was that lady born without shoes?"

As refreshing as those moments were, they were unreliable.

Motivation was usually my responsibility.

"Level, is spelled the same forward and backwards. Those on the upper level can always hit the bottom, and those on the bottom can always rank to the top. Envision your footprints up there already trailing, and your feet will soon follow suit." — Anthony Liccione

I had been invited to join an Anthony Robbins seminar, Awaken the Giant Within, a few years previously. It had been a life-changing experience and at the end of the weekend I, together with thousands of other participants, had walked barefoot over red hot coals.

I had fire walked.

I had also been introduced to the power of the mind and the benefits of getting it to work with you. I was already a strong advocate of minding my language, using positive words and phrases whenever I could, or could remember to. Visualisation was an even stronger tool to employ.

I habitually imagine being covered by a healing and protective crystal dome before going into social situations or crowds of any sort. Alternatively I visualise a big bear or eagle cloak over my shoulders as a form of protection against negativity and even germs and bugs, as the Shamans do. It works for me.

And that's the only validation I need.

I had been taught these specific methods but was keen to develop them for my personal circumstances.

When running up hill I would 'tell' myself that I was running on the flat. It's quite alarming how effective it is when you do it well. We are easily duped!

I recall a meditation led by the Dalai Lama in which we firstly imagined the body hot, then cold, then heavy, then light. It is a powerful tool for training the mind.

One of the most useful visualisations I have found when encountering a specific pain when running is the 'big fish little fish' (no cardboard box) one.

Running the London Marathon (in trainers), I had an intense burning pain in my left thigh about twelve miles in. Imagining little red angry fish at the site of the pain for a few minutes, I then visualised big jolly blue fish bobbing along and eating up all the little angry red ones. It works every time. It is also a pretty effective pain relief process in other situations too. I used it a lot during training and also during the barefoot run itself. After about six or seven miles I had a pain spasm on the top of my right foot; the 'big fish little fish' didn't relieve it totally but I am sure the little fishy fellas played their part in me being able to finish the run.

Another trick I have is to actually breathe into the pain; imagine your breath being taken straight to the site of the niggle and dissolving it. Some people prefer to visualise the breath actually entering the body directly at the site of the pain. It really is a matter of employing whatever works for you. And believing in its effectiveness.

Another fabulously effective aid to running barefoot (or otherwise I'm sure) which I invented whilst attempting to conquer some of our big Welsh hills is a method I call running like a river.

I practice the Qi Gong discipline of Zhan Zhuang or standing like a tree, to increase energy and free emotional blockages. I adore it.

In this day and age it is a tricky concept for most to hold a single position for up to an hour before changing to a similar pose for the next hour. Of course it doesn't have to be an hour, five minutes is better than nothing at all.

During one Zhan Zhuang class we were shown how to 'move' energy gently with our hands. It was this that sprung to mind early one morning as I contemplated another steep hill run.

It was worth a go.

I simply imagined scooping energy from around my body and from the universe and guiding it into my Dantien (belly button region) to provide me with the energy I needed to run up the hill. Then I imagined a thick silver thread coming down from the top of the hill to the bottom. I made overlapping hand movements as if climbing up a rope. And thirdly I used my hands to 'pull' myself through the universal energy like paddles, pushing the energy behind me and therefore propelling myself forward, a little like a subdued front crawl.

Whether or not these actual visualisations and efforts made the difference or whether it is was simply keeping my mind off the steepness and severity of the hill I'll never know, but I do know it worked and still does.

You can't worry too much what you look like though. And the good thing is that when you are running up the side of a rugged, remote Welsh mountain, you don't have to.

31.
Knocking on heaven's door

'I discovered that I had been plotting and planning with the legend that is Eric Clapton.'

I had found my childhood confusing and my teenage years challenging but my early twenties were to provide the steepest learning curve of all.

In a very short time, my knight in shining armour had indeed lost his shine. But not before I misplaced my very last remnants of self-worth and self-respect. Funny how it works like that. Like a dripping tap, any little crust of self-preservation I had managed to retain was soon eroded by cruel taunts and verbal abuse. I was being bullied and berated for having the very same attributes that he had found attractive in the first place.

I reeled emotionally as I tried to make sense of it all. I tried to retreat but he would entice me back with abject apologies, generous gifts and cleverly administered kindness. I slowly let my own identity and life slip away. I was exposed. Vulnerable. And under attack. Soon my only sanctuary was provided by the very person who was attacking me.

Unable to make any sense of the situation, I had hidden my confusion, the bruises and my fear and made the monumental decision of moving to the little, undeveloped Caribbean island of Montserrat with him. He was going to take twelve months out. Life had been stressful, business had been good.

My own business had been good too and I loved it. I loved the camaraderie on site, I loved being outdoors and yet I loved the idea of exploring the big wide world even more. My brother had recently qualified as a mechanical engineer and decided to take a break from the spanners to work with Dad in my absence. It would be a change for him. It would be a change for us all.

One month I was working on a building site, mixing cement and concrete, driving a transit van and living in a mobile home; the next month I wasn't working at all. I was mixing with celebrities and millionaires, driving a brand new Toyota double cab Hilux and living in a luxury villa overlooking the Caribbean Sea. I had gone from ordering sand from the builder's merchant to lying on it in the sun.

Of course it was idyllic. And traumatic. I was out of my comfort zone. And that was an understatement. More than ever, I needed to be busy to cope with this emotional disorientation, to give me a familiar point of reference.

I maintained my column for the Tindle Group of newspapers. I would write a 'what to do in your garden this week' style column on a boat, plane or beach and then find somewhere to fax it back to the UK.

It seemed a perfectly normal thing to do and allowed me a little reassurance by retaining an identity I recognised. As I often used the local newspaper, the Montserrat Reporter's, office to fax from, I became acquainted with the editor and was soon asked to contribute to the island's magazine.

Montserrat is one of the lesser inhabited islands in the Caribbean. With no harbour or jetty, it wasn't geared to take cruise ships and, being a volcanic island, it also had black sandy beaches which were not as attractive as the more traditional tourist-type white sands of neighbouring and more accessible islands. Tourists and news were thin on the ground. I inevitably spent a lot of time trying to get suitable material from the locals. It proved a great way to get to know people on the island.

When we arrived, the Montserratians were still rebuilding their lives and properties after the devastating effects of Hurricane Hugo which had rendered 90% of the 11,000 inhabitants homeless in 1989. The jetty at the capital, Plymouth, had been destroyed and it had 'an island that time had forgot' flavour. As we flew in for the very first time, the pilot of our small four seater chartered plane told us, "there is a time difference here; you may want to put your watches back 100 years."

The runway was rustic. I had created and maintained better looking lawns. It was also flanked by small-plane debris.

Because of its sleepy mañana attitude, Montserrat proved popular with 'snowbirds' - those who travel south for the winter to their holiday villas in the sun. I loved them. They were retired, rich and in turn, loved an 'English' gardener. I was soon giving design advice to other 'snowbirds' on the island and learning a lot about tropical plants and volcanic soil.

I learned that the cashew nut grows on the base of a cashew apple. To the locals the apple is more prized than the nut, for its thirst quenching qualities. I grew peanuts - underground of course and discovered that a 'banana tree' is actually a shrub; once the plant has fruited it needs to be cut hard back in order for new shoots and fruits to form. I became a dab hand at harvesting and machete-ing the tops off of fresh coconuts for a refreshing drink.

I also sent a whole coconut to my mother who was still in hospital. It seemed like a good idea. Still encased in its waxy outer shell, I simply wrote the address on the actual coconut, had it stamped and put it in the post. It had been with the intention of speeding up her recovery although it nearly had the opposite effect as the psychiatrist challenged her excited claims of receiving a real coconut in the post.

And of course there was the fishing. Big game fishing. We fished for marlin, tuna, dorado and even sharks at night. We would get a bucket of animal blood and guts from one of the local farmers, freeze it in one of the local supermarket's large chest freezers and then tie it behind the boat on a long rope at night. As the frozen offal thawed, it created a blood slick on the surface of the water which would attract the sharks.

A new, enthusiastic, wealthy 'resident' who loved fishing was a big bonus for the local fishermen and they would eagerly trade their local knowledge for outings in our well-equipped boat. I learned how to find and fish the 'drop offs', a place where the fish would hang out. I also learned not all fish were safe to eat.

Unfortunately I encountered Ciguatera soon after we moved to Montserrat. Ciguatera is a type of poisoning caught from eating a certain fish. In my case, I had eaten a flat fish called a Horse Eyed Jack which we had bought from an unscrupulous and short sighted fisherman on the beach. The Jacks feed on pelagic fish which have fed on dead coral and are renowned for their toxic qualities. Usually a little of the flesh will be given to a chicken as it can't vomit; if the chicken dies, the fish is rejected - or sold to an unsuspecting Welsh woman.

I was incredibly ill; found unconscious on the villa floor one evening and hospitalised for three weeks. During the first week I had no feeling in my legs whatsoever. I was continually vomiting and had high blood pressure. A doctor was flown in from the medical school on the neighbouring island of Grenada and told me that there was no cure; the symptoms would just ease.

Ciguatera affects the nervous system chronically; therefore I could expect ongoing problems with numbness and or tingling, myalgia and joint pain. He added that, on the bright side, I was lucky. Some people with such a severe reaction die. Of course I might too but it was unlikely now as I had survived the first week.

I woke up one night with a group of gospel singers surrounding my bed, singing and chanting. I thought I had died and was in heaven but maybe it was only their prayers that stopped that thought becoming a reality.

The doctor also warned me not to eat fish or shellfish again as it could trigger the toxins that would remain in my body. Caffeine and alcohol could have the same effect but no one knew why. These triggers could result in a return to partial paralysis and there was no telling where it could occur, physically, as well as geographically! Avoiding eating fish in the Caribbean was challenging but I had good incentive.

Over dinner one night when I had to yet again explain my refusal of a fish dish and alcohol, the 'snowbird host' turned out to be a medical consultant from New York. He had some experience of Ciguatera and advised that I introduce fish back into my diet very slowly to avoid unexpected ingestion causing a problem later. He confirmed that there was little understood about the poisoning and its numerous side effects but vigilance was the key.

I could do vigilance. I was used to keeping an eye open for an opportunity.

Whilst writing for the Montserrat Reporter, I discovered there were several horses on the island which had been turned out into the hills after an attempt at establishing a riding school had failed. Over a number of years, they had become feral and elusive.

After eventually locating the herd, I visited a beach bar on returning from my efforts to get near the horses one lunch time. I was extremely hot and a little frustrated after pursing them on foot for hours. A quietly spoken guy commented on my disheveled demeanour and was curious about the cause. After explaining my exploits and my intention, he revealed that he spent quite a lot of time on the island and was also looking for an interest whilst he was there.

He also loved horses and riding.

He inferred that he also had a bit of cash which he would gladly contribute if I needed it. We met several times and outlined a plan to re-establish a riding school on the island. He introduced himself as Eric and that's as much as I knew about him. He went back to the UK leaving me to push on with the plans. In his absence I discovered that I had been plotting and planning with the legend that is Eric Clapton. I had been Knocking on Heavens Door again but in a much more pleasurable way.

Paying a little more attention to the people I met on the island, I subsequently recognised Midge Ure. And Sting. And Mark Knopfler.

Montserrat was home to the famous Air Studios and despite the fact it had closed a little while ago, many of the musicians still had homes on the island and still appreciated its low key status.

Mark Knopfler used to wind surf with Danny, a popular local who had become our right-hand fisherman. The record producer and owner of Air Studios, George Martin, had bought a little wooden boat for Danny who had also partnered a lovely blonde English girl called Margaret.

We got on famously. Danny was delighted to share the fact that when Dire Straits had recorded the track 'Walk of Life', Mark had insisted on adding 'Danny do the walk of life' to one of the verses. We all spent many a night quite literally 'twisting by the pool' of George Martin's villa.

As well as being a guest at homes that resembled hotels, I was fast becoming the 'Hostess with the Mostest'. Always keen to keep busy and to satiate a thirst for broadening my horizons, I dusted off the knowledge I had absorbed whilst waitressing at Glan y Dwr, the Gourmet Restaurant. I had learned about fine wines, even finer food and how to prepare and present them both. People didn't have to know how I knew; I just knew.

I became a bit of a novelty; a little seven stone, blonde Welsh girl who could not only wield a mattock in the yard and fight big game fish but who could also turn the plantain she grew and the fish she caught into a delicious dinner.

I was very grateful for my practical abilities. They outstripped my emotional ones.

This chapter of my life still evokes some of the happiest memories and some of the most painful ones. I was still being beaten up on an 'unreasonably regular' basis and my partner had also confiscated my passport after I had foolishly threatened to leave.

And yet the time between the bruises was exciting and endearing. A combination of remorse and shame resulted in me being revered and lovingly indulged. With endless funds available and limited facilities on Montserrat, we travelled a lot. Regular chartered plane trips to the neighbouring islands of St Martin and St Kitts or to stock up on fishing tackle at Antigua, two weeks in the luxurious five star Cap Juluca Hotel and Resort on Anguilla, another two weeks big game fishing off the coast of Venezuela, three weeks on the remote Bird Island in the Seychelles where I spent time wandering with giant turtles, spotting the endangered magpie robin and staring in wonderment at the dragon trees which actually 'bled' when cut.

Whilst there we fished hard out of a tiny wooden boat with the islander's best fishermen and kept the locals supplied with fresh fish. We had brought new rods, reels, lines and lures. We became heroes and were treated like royalty.

The sea was alive with magnificent and majestic sail fish. They would 'stand' and walk on the water when hooked, in the sunlight. I was in complete awe of nature on a totally new level.

These times of absolute wonderment were magical and addictive, being sandwiched between fear and pain of the emotional and physical type. All my pleasure and sanctuary continued to be provided by nature and animals. I justified catching fish as we were feeding the islanders; it wasn't for sport, it was for survival.

Big game fishing is also a powerful contest. The fighting fish would often outwit and out play us. More often than not, the fish won.

"That's why it's called fishing and not catching," one gnarled, wise old fisherman enlightened me.

And it was not catching that resulted in me being held over the side of a boat in the middle of the Indian Ocean, threatened with being thrown in and left for dead.

And all because I had lost a black marlin.

They are probably the most prized catch amongst big game fishermen because of their speed, agility and wiles. They are notorious for accelerating past the boat in an attempt to throw the hook, which is precisely what this particular one did. Getting away with his life nearly cost me mine.

Our extravagant adventures included chartering a 90 ft Hatteras and crew in Florida and fishing the Florida Keys. While there, we also stayed at the luxurious Magic Kingdom Hotel at the Walt Disney World Resort. As well as shaking hands with Mickey Mouse, I was shaken over the third floor balcony of the hotel by my ankles, after another violent outburst.

From magic to misery in moments.

My parents had no idea of my predicament. I wasn't in touch very often. I had been labeled a difficult child, a terrible teenager and it was presumed that I was now being an aloof adult.

I missed home dreadfully.

I sometimes wonder if it was my experience of dealing with the extremes of behaviour associated with Mum's illness that enabled me to be accepting of such violent extremes within a relationship - or did my expectations evoke it?

I know that during times that were darker than black, I would always retreat into nature. And in the brightest of times nature would make up the audience for sure. I also had tremendous and continued belief in the elementals and know, without doubt, that the nature spirits kept me safe.

I retained my faith in them and the power of something far greater than myself and, on the few occasions where I have thought I was going to die, I have accepted the situation gracefully.

I was also barefoot for a lot of that time.

32.
The route of all evil

'My hormones were possibly being
as unhelpful as my Chimps.'

The great Jim Morrison said, "I think the highest and lowest points are the important ones. Anything else is just... in between."

I was still training hard for my attempt to run across Wales barefoot and I didn't seem to spend much time in between.

I seemed to either be running high or sitting with my head in my hands, full of doubt and darkness. How on earth was I going to run across a country in a weekend, let alone without shoes? Have you seen the size of those mountains? I didn't even have a route planned.

It was all proving to be too much. It was mid-February. Work was hard. Running was hard. Life was hard. What was I doing? This was all self-imposed. I berated and beat myself with that big emotional stick that I carry around continually.

My legs and feet had taken on a complete identity all of their own, often refusing to liaise with the rest of me. They ached, they sparkled, they danced, they sulked, they investigated, they refused, they played, they protested. I tried to understand the fluctuations and manage them. One day I was full of energy, enthusiasm and definitely on track. The next I was overcome with doubt, loneliness, exhaustion and felt defeated.

It was proving to be a long, lonely road.

On a good run, I felt like I was running on air. I had a Tigger-type bounce and a Winnie-the-Pooh style wisdom. I ran so lightly and quietly that I actually crept up on birds feeding in the verges, startling them. It was funny. I loved it. 'The wonderful things about Tiggers...'

Other times I felt like Eeyore.

I tried to be nice to myself; understandably I was tired. I told myself to rest. To be kind to myself. To have a hot bath. To do something nice. To curl up in front of the fire and read.

I couldn't.

I couldn't keep still. Physically or mentally.

I read The Chimp's Paradox by Dr Steve Peter's. He has guided many successful athletes to combat their negativity and tendency to self-sabotage. The Buddhists call it the 'Monkey Mind'. Either way, it seemed that my inner chimp or monkey mind couldn't relax either. I soon reverted to berating myself. It was never very long before I was losing patience with my aching body and apprehensive mind, "For Christ's Sake, you only ran 8 miles; c'mon, get a grip. You want to do this? Then buck up ..." My chimps and monkeys chattered and chivvied incessantly.

I wondered how much of it was hormonal. The more aware I was becoming of my fitness levels, moods, foods, recovery times and other personal aspects of the training, the more it seemed that my hormones were possibly being as unhelpful as my Chimps.

There was definitely a pattern. A few days before my period, I would feel sluggish, lethargic and all out of sorts, then I would have the most excruciating period pains (for which I stubbornly refused to take pain killers) and which were often painful enough to stop me walking let alone run. I would also feel bloated and just generally weak and feeble. If that wasn't enough to deal with, keeping a diary showed that I had an extra dip mid cycle where I would feel tearful, isolated and quite depressed. A double-dip depression. I calculated all in all that out of every month I could write off ten days because of hormonal horrors.

That's a lot. Over six months, its sixty days or two months. That took my training time down to four months.

I had to factor in these unmotivating moods and stop being so hard on myself. Kindness was the way forward.

Thankfully, other people aren't as hard on me as I am.

Mike always had words of encouragement. "Nothing's impossible," he assured me. As I tried to explain a new type of pain, he shrugged. "That's good. Pain is weakness leaving the body. Lance Armstrong says so."

Mike also had some well-timed words of wisdom as I shared my woes at having heard a few people being damning and derogatory about my challenge. It had wrong-footed me slightly.

"Don't worry about Team Twat," Mike instructed. "That team always shows up; they don't do anything themselves but don't want to see others doing anything either. They are a big team but they have no power unless you give it to them."

Mike also introduced me to a foam roller, to ease out my knotted muscles. "This is going to be your new best friend. Borrow mine until you get one of your own. Oh, and here's a golf ball," he added, throwing it across the gym for me.

Rolling my bare feet over a golf ball eased out the tight muscles and ligaments in the soles of my feet. Muscles and ligaments that weren't used to being engaged whilst in trainers and shoes, ligaments I didn't even know I had.

I hadn't planned on aching so much. The view through my rose-tinted spectacles had very much been of myself running fleet-footed through the Welsh countryside like a Taramahara crossed with Julie Andrews - with even possibly a bit of Bambi and a few Disney-esque butterflies thrown in. I was more like a cross between Julie Walters and the donkey from Shrek and thinking of throwing the towel in.

I despaired. I had investigated (and blamed) every possible aspect I could think of, from poor nutrition (but George Cooper had sorted that out) and a lack of training savvy (but Mike James and Michelle Blackmore had that covered) to the incessant wet weather and even the alignment of the planets and the fact that Mercury may just have been retrograde.

But still I jogged on. The bad days were spent dragging my feet and arguing with the inane chatter of my internal primates and on the good days I got out and ran like Zola Budd, lifting my knees, landing on the fore front of my foot and looking forward to my fish soup.

I was also becoming obsessed with running surfaces and especially road surfaces. Everywhere I went, whether I was driving, walking or running, I studied the road surface, assessing it for its suitability for running on with bare feet. I dreamt about black ribbons of tarmac weaving their way though valleys and over mountains. I composed poetry in my head, 'An Ode to Tarmac', as I ran. It was never-ending.

The other thing that remained constant was the tremendous support and encouragement I received. It became apparent that there is actually quite a sizeable hidden tribe of barefoot runners permeating the country and indeed the world.

Even the legendary Daniel Lieberman emailed, "Good luck Lynne and have fun! Run lightly and gently!" There was huge support via social media. Several barefoot groups had taken my challenge under their wing and I received a lot of well-meaning advice and good wishes from people I had never met and may never meet.

Nonetheless, I could rely on their support and encouragement.

33.
Homeward bound

'.... and the clothes I wore which were now blood stained.'

Friends I had made on Montserrat once witnessed the tail end of an abusive attack I received and insisted I return with them to their ranch in Connecticut, just for a break. Or, as they warned to avoid a break.

Sue was a famous event rider with extensive stables and her husband Nick had a large and successful landscaping business. In theory it was a dream situation. I loved the yard work and rode some of the most beautiful horses I had ever seen. I also worked with Nick and was soon driving around on my own overseeing the Portuguese labourers on various sites.

Sue and Nick offered me a beautiful apartment, a generous wage and a Green Card. I refused and went back to the Caribbean and the chaos that had become my personal life. It was what I knew and more distressingly it was what I thought I deserved. I was worried Sue and Nick would find out what I was really like. It is a fear that I have crossed paths with many times since, particularly in relationships. And yet I still don't know exactly what it is that anyone would find out. It seems to be wrapped up in the fear of letting people down or disappointing them. Being disappointed was far easier to deal with than the risk of being considered disappointing.

I still consider my time in the States to have been one of the most civilised and comfortable times of my life and yet I still self-sabotaged. Even so, I am pleased to say that the only regret I have about leaving Sue and Nick's haven, is that we eventually lost contact.

Soon after I returned to Montserrat, the volcano Chances Peak began to reawaken and stretch. It had been dormant for over 300 years and had never been considered to pose any threat or problem. I had spent a lot of time with the wild horses at the base in the Soufrier Hills and walked up into the volcano crater itself many times. Despite a strong sulphuric smell of rotten eggs and large Iguanas lazing on the hot rocks, there wasn't much to write home about.

A small earth tremor persuaded us to sell up swiftly and just six months after we had left the island, the volcano erupted with intent.

Despite evacuations to the north of the island, which was totally undeveloped, twenty people died as they naively believed the volcano was just angry. The south of the island was completely buried in hot volcanic ash and only the tip of the Clock Steeple was visible.

My Montserratian episode had come to a close. I often think about the island and the islanders and would dearly love to revisit the place where I morphed into maturity.

We left the Caribbean island for Ireland, selling a cliff top villa and buying a country estate in Tallow, County Waterford.

The transition was made via London where business matters were brought up to date. We spent two weeks in the city. It was a horrendous shock to my system after the familiar, friendly, nature-soaked, Caribbean island. We stayed at The Ritz. For two weeks. In a suite. Far from feeling like a luxury, apart from the relative peace and quiet it offered, I found it hard. The perfection that permeated everything felt oppressive. I asked staff please not to make the room up, it would give me something to do, to make the bed. I couldn't bear everything being in exactly the same place when I returned every day. I befriended the door men and other staff in an attempt to normalise such a pretentious environment. I learned about their own family life and missed mine even more.

We hosted business dinners in the suite; there was endless champagne, lobster, steak, brandies, referred to by letters like XO, balanced over coffee cups and fat cigars served amid a sea of silver and quaffed and consumed with no regard for cost. I still couldn't refrain from totting everything up in my head, calculating the cost and equating it to a wage.

Tweeds and shooting paraphernalia were purchased and replaced shorts and fishing lures. Fly rods replaced fighting belts. It was like being Mr. Ben and creating a whole new identity. Only it wasn't going to be for a day. It would be a whole new chapter.

Once ensconced in the leafy Irish countryside and eager to be working, I became acquainted with a well known local race horse trainer Bobby McCarthy and was soon providing livery for Robert Sangster's brood mares. It was another incredulous situation to be in. Yet again, it was my time spent in the field, rather than in school, which proved the most useful. I re-engaged my knowledge from the National Hunt Yard and once again the animals helped ease the pain of the mental and physical abuse that continued.

Mum was still in and out of hospital and although now much nearer to Wales, I was still being prevented from having much contact with my family and only allowed to call home when my partner was in the same room.

The violence would increase in direct proportion with my vocalised desire to return home and so I put it out of my mind.

The horses became more than allies. They now took the blame for my inability to visit home as well as for my bruises.

The mental abuse was proving harder to deal with than the physical. Constantly being told that I was useless, stupid and that I would end up back in the gutter if I ever left. And, of course, no one else would ever want, or put up with me.

I believed him. 100%.

Thankfully not everyone did.

One day while walking though Cork, I was approached by a smartly dressed and well-groomed man who asked if I was interested in working for a Model Agency in the city, called Profile.

I was a tiny size six due to a combination of worry and work and dressed in black. I always wore black. I had way more common sense than fashion sense and black was easy. Black went with black. Unbeknown to me, all the Profile girls wore black. It was what had caught Ed Jordan's eye. He owned Profile Model Agency and, over a coffee, offered me a job there and then. No contract, no fees and absolutely no correlation to the life I knew.

I had no horses in livery at the time and nothing to lose.

I credit my continued good fortune to one of two things: either creating an opportunity or saying 'yes' when one is offered.

I said, "Yes."

My partner had to go to Europe on business for a couple of months and wanted to travel alone. We decided that we would get a farm manager and his wife to move in and continue the running of the estate. I chose not to go home in favour of going to work in Cork. I had too much to explain if I went home. It was easier this way. Easier to go forward than back. Somehow. It made sense at the time. Or as much sense as anything could under the circumstances.

I moved in to a flat in Cork with three other Profile models. It had all the hallmarks of being a living hell for me - living in a flat in a city centre with three other girls, working with people and fashion, no animals, wearing make-up and socialising substantially. And yet I did it.

And I loved it.

I loved the combination of earning and learning something new - how to work the cat walk, even the novelty of make-up. It was like putting on a mask. Once I had my make up on it was show time.

A little ironically, one thing I couldn't master were high heels. Ed devised a simple solution. I would grace the cat walk in my bare feet; it would be quirky and he felt 'it suited me'.

Another clue sent by the universe and spectacularly missed by me.

After the Fashion Shows we would go to a Night Club and be treated like royalty. We were the Profile Girls. We would sashay and strut past all the queues and be shown straight into the VIP lounges. It was frivolous, fun and fabulous. Myself and one other girl didn't drink alcohol and gelled immediately. Majella was a farmer's daughter who was as unfamiliar and bemused with the whole experience as I was. We agreed we would just enjoy it for long as it lasted. We were both sure we would be 'found out' eventually.

I was given the additional responsibility of being the spokesperson for the agency and did numerous interviews for TV and radio. I got more and more involved with the running of the agency and the organising of the shows, managing the girls and thoroughly enjoyed it all.

For the first time in my life I had more people than animals and plants in my life. I was in a flat in Cork. We didn't even cook; we always ate out. It was about as surreal as numerous other episodes in my life and I simply embraced it by acting my way through it all. Acting by-passed the need to adapt and all the emotional connotations and painful realisations that would have to have been endured by being myself.

I had become worthy of an Oscar.

While at Profile I applied to be considered for an IQ appraisal via a small Mensa test in a newspaper. I had been accused of being thick and stupid a lot during my life and an IQ test seemed an obvious way to find out if I was.

I had a letter congratulating me on the result of the brief application test and inviting me to sit a supervised appraisal at a choice of venues.

There was one in Cork.

I sat it. There were only two of us and a supervisor. I thoroughly enjoyed it. The questions provided a stimulating and satisfying mental workout. It was exhilarating. I would have to wait for my official Mensa score to see if my enthusiasm reflected my ability.

The paradox is that despite my emotional insecurities and battered self-esteem, it never ever, not for one nano-second, occurred to me that I may score low.

I didn't.

My IQ was 155. I was in the top 1% of the population and was officially a member of Mensa.

Having a high intelligent quotient obviously is no indicator of being smart. When my partner returned from Europe, I joined him back at the estate and turned my back on yet another civilised and comfortable environment and episode of my life.

I felt so much better about myself. Surely our relationship would only benefit from that? I was so naive. My increased confidence only increased my partner's anger. Life reverted from terrific to terrifying with alarming speed.

In the first week of returning back to the estate, two of my kittens died with inexplicable injuries. I had also just adopted a beautiful little terrier from a local farmer. The dog was blamed despite the fact that he had been with me all day and had absolutely no blood on him whatsoever. There had been a lot of blood. Alarm bells were starting to ring.

Days later I spent the longest hour of my life being held against a kitchen wall with a loaded shotgun at my head. He kept flicking the safety catch on and off as he lurched between threats and accusations.

I thought I was going to die. And not for the first time. I prayed hard. It wasn't the first time for that either.

Eventually he got bored and took the barrels that had been pressed against my cheek out to shoot rabbits.

Still shaking with fear, I made my through the long corridors of the house in a daze. I found what I was looking for with little effort. I hoped it would act as a sort of insurance policy. With my legs threatening to buckle beneath me, I managed to get to my truck and push it under the seat cover.

As I slid down the side of the truck and crumpled in a heap on the gravel, I heard his feet scrunching toward me. Without hesitation he strode past, apologised for his behaviour and suggested that we went to my favourite seafood restaurant in Youghal for dinner.

The next day I came back into the yard from the stables and immediately saw a lump of bloodied fur by the back door. My heart banged, my own blood raced, I screamed - it was my dog.

I thought he was dead.

I scooped him up and found that he was still breathing, albeit shallowly. I wrapped him in my coat, put him on the front seat of my truck and drove out of the long tree-lined drive.

I drove to the vets. It transpired that his injuries were remarkably fewer than first thought. He had a deep cut on the side of his jaw was missing two teeth and had numerous grazes on his body. He was also obviously very bruised and was in shock.

The vet told me the injuries were synonymous with being kicked violently, possibly by a horse. We had no horses at the time.

I stayed at the vets while they cleaned and stitched him up. I was distraught. The veterinary nurse had commented on my own bruises and I completely broke down. I had nothing left. I wasn't going to act my way out of this.

I told them as little as I could get away with sharing. They gently suggested that they find my dog a new home and that I should return home to Wales. The veterinary nurse said she knew the perfect family who would adore the dog. They had just lost theirs after fifteen years. There were two young children. She made a call. They had been delighted to be able to help, they would wait until he had recovered and they would take him. They insisted the nurse pass on their details and I could call or visit anytime. The vet added there would be no charge for the treatment.

I wept uncontrollably and I promised I would return to Wales.

I drove through the night to Rosslare. I had no money, no passport and no ID. Just a truck and the clothes I wore which were now blood stained.

And the item I had still under the seat cover.

As I sat in the truck staring blankly at the ship, a ferry worker approached me. I sobbed my sorry story and explained I had no idea how I was going to get home.

He told me he had a daughter of his own and waved me onto the ferry.

I was in no fit state to go straight home. I called a friend who lived about twenty miles away from my parents; she was away for a fortnight but she told me where the house key was, where I could find some cash and that I was to stay there as long as I wanted. It was remote. It was perfect. I stayed there for four days and slept for three of them.

I called my parents to tell them I would be home in a couple of days. Reverting to actress mode, I explained that a friend needed some help moving horses and other animal stuff so I would be staying with her en route. I would be at her cottage until I got home. I gave Dad the number. There was no mistaking the relief in his voice.

Mum was back in hospital. Business was good. My brother was fine. Nothing much had changed. They would see me soon.

I hadn't seen them for two years.

The phone rang later that evening and I picked it up expecting a call for or from my friend or maybe to hear my Dad.

My heart stopped as a familiar menacing voice screamed down the phone threatening to kill me and my family. He would "hunt me down and finish me off properly for what I had done to him. How dare I leave him? After all he had done for me. I would regret it."

I can vividly recall the view from the little cottage window across to the mighty Pen y Fan mountain range. As if in slow motion, I calmly told him about the item I had taken as an 'insurance policy'. He screamed that I was lying.

I suggested, in the same flat tone, that he went to check and then call me back if he wished to do so. He never did. I haven't heard from him to this day.

It was over.

It had been the hardest time of my life. Surely it could only get better? Couldn't it?

34.
Massif challenge

'Surely someone had blazed an historical trail which I could follow.'

While initially planning my epic barefoot attempt, I considered running it on my birthday. My 47th birthday. For no reason other than it seemed a nice way of marking a birthday. But my birthday is in the middle of March and apart from restricting my training time considerably, I wasn't confident that the weather would be on my side then either. Instead my birthday was spent travelling up to mid-Wales to recce the route. It was just six weeks before my run and, naively, my very first recce. Contrary to most people's idea of a birthday celebration it was a particularly sobering event.

At this stage I was still planning to run the route used by the Walk Across Wales which started at a tiny hamlet called Anchor on the Welsh-English border and ended at Clarach Bay on the west coast of Wales. It also incorporated the highest peak in the Cambrian Mountains, Plynlimon at 2,467 ft., and a massif that dominates the countryside of northern Ceredigion. In geology, a massif is a section of a planet's crust that is demarcated by faults or flexures. In the movement of the crust, a massif tends to retain its internal structure while being displaced as a whole. The term is also used to refer to a mountain formed by such a structure. Plynlimon is also the source of the longest river in Britain, the River Severn.

Folklore also says there is a sleeping giant in Plynlimon.

Careful consideration of this route had also shown that there was no recognised or marked path across the mountain range – it was a case of finding your own way.

That flagged up a possible stumbling block straight away. Navigation is not my strong point and whilst I relished the idea of putting myself through a crash course in orienteering, there simply wasn't the time. So it made sense to recce this part of the route, at the very least.

Daunting doesn't begin to describe it.

It was a grey spring day and the mist hung low over the peak of Plynlimon, like a duvet over the sleeping giant.

I stood at the foothill of the great mountain and felt completely and utterly deflated.

I had already got lost whilst looking for the mountain. All 2,467 ft of it. In a car. With a map.

And now I couldn't even see the top. There were no evident paths anywhere, not even a promising sheep track and the terrain was the most inhospitable tufty grass hiding equally inhospitable ankle twisting crevices.

Bugger.

I made a Native American style offering of white sage and tobacco to the veiled mountain and asked respectfully that the route be revealed and the journey unfold as it was supposed to. I gave gratitude and I left.

I drove on to Clarach Bay, the place where I was planning to end my barefoot run. It was on the coast, on the sand, by the sea; surely that part would be satisfactory and straight forward.

It wasn't.

If I had had boots on, my heart would have sunk into them. I don't want to be disrespectful about any part of my beautiful country but suffice to say that the grey, stony beach lazing behind a garish amusement and caravan park did not fill me with joy. Not even close. It was not an ideal destination; not somewhere that I would run joyfully toward, not somewhere I would look forward to finishing my epic run.

Bugger.

Sitting in the truck as the rain lashed against the windscreen, I couldn't even see where the sea joined the sky, and it all just merged into a depressing grey mist.

Studying the map for an alternative destination, a number of options were ruled out immediately on technicalities. I moseyed on up the coast.

Borth, the next village, didn't really appeal either; it just didn't feel right, but then a few miles further north, lay Ynyslas. It was sandy and expansive and felt good. Surely this would be a better choice. I just liked the feel of the area. It was welcoming despite the grey curtain of mist. Yes, I liked this place.

And I soon discovered why.

It transpired that Ynyslas is also a Nature Reserve and the site of an Ancient Forest. Pieces of petrified wood can still be seen on the beach at low tide. Of course I would want to run toward such a place. It was inspiring.

So I had found the finishing point. But of course that would affect the rest of the route.

Running out of daylight and energy, I turned for home and decided to detour through Llanidloes to drown my melancholy with a cup of black coffee. Whilst looking for a cafe, I popped into a book shop to a look for another possible meaningful route across Wales. Surely someone had blazed an historical trail which I could follow.

There were the routes I knew, Offa's Dyke and Glyn Dwr's Way, but they, nor anything else, were remotely suitable. Commenting on my bare feet, the owner asked what I was looking for - something 'unusual' she suspected.

Hearing my plans, she explained that there was another annual organised walk across Wales held by the Rotary. The organiser had a shop a few doors down the street.

Coffee would have to wait.

I love the way the universe nudges, tugs and guides you in the appropriate directions. I think of it as the original, organic and delightful Sat Nav.

Following a quick chat with 'Mr. Rotary' who was totally unfazed by my intention and far more interested in closing his shop on time, I had a new route option. He kindly agreed to email the route they used. It wasn't dissimilar to the Walk Across Wales which I had already researched, apart from the fact they walked in the opposite direction, away from the coast and actually started the walk at the end of an estuary rather than the coast. This meant that they also omitted the mighty Plynlimon. They let the giant slumber.

There were pros and cons. It would all have to be thought through and considered. But at least I had another option.

A few days later I laid out the three vast Ordinance Survey maps that covered the possible routes, they also covered the entire floor and sofa of my living room.

Good grief, it was a long way.

I decided there was absolutely no way that I could run away from the sea inland toward Anchor. It just didn't feel right.

Neither could I justify my run across Wales starting six miles inland at the end of an estuary.

To compensate, it meant that my 45 mile run would have to be a 52 mile run. Two marathons, back to back.

However, the big bonus of the Rotary Route versus the Walk Across Wales's route was that I avoided Plynlimon. I recalled how the magnificent mountain had hidden under the grey misty blanket - as though indicating that was not the right route.

Intuition = tuition from within. I silently gave thanks to the mighty mountain once again.

So all I had to do now was reverse the Rotary Route and find a suitable route to take me the last six miles into the beach at Ynyslas. I lit the wood burner and rearranged my map carpet.

Overwhelm doesn't even begin to cover it but sprinkled over that ocean of overwhelm were the tiniest sparkles of adrenaline and excitement. I sharpened my pencil, gathered my highlighting pens, opened my notebook, shuffled my sheets of Rotary Walk directions and took a deep breath.

And began writing.

35.
Dying to get it right

'He sighed, put his head in his hands and said, "I have found a lump under my arm".'

One afternoon, not long after returning home after two rather surreal years abroad, which my parents knew next to nothing about, I sat on the lawn with Dad having a cuppa. We had been working together all morning in his garden and I knew there was something on his mind.

It was good to be back; very little had been spoken about my travels. It was 1993 and I was 26 years old. So much had changed and yet some things hadn't changed at all. My brother was enjoying the landscaping work and business was good.

I had rented a cottage near to my parents' and started helping out with the business now and then. I was still incredibly raw emotionally and spent a lot of time walking and wondering about life in general.

Dad was also preoccupied and it wasn't with Mum or work. Mum was back in hospital, having been sectioned again several weeks before I got home. Dad was back on the daily visits but it was a familiar and non-taxing routine. I kept pushing him. "What's wrong?"

I knew something was. I told him I would keep on until he told me.

He eventually sighed, put his head in his hands and said, "I have found a lump under my arm."

He went on to explain, reluctantly, that he had had it for 'probably a couple of months the way time goes'.

'No it didn't hurt.'

'Yes it had got bigger. '

'No, of course he hadn't been to the doctors. Well, only about his thumb last year.'

He added quietly, "I know its cancer."

I ignored his last sentence.

"What was the matter with your thumb?" I asked.

We were both just staring straight ahead. I pushed my coffee cup into the grass trying not to spill the contents.

It transpired he had had a thick black vertical line on his thumbnail. He had initially gone to the doctors because a black thorn had gone septic in his arm. He had mentioned his thumb whilst there. The doctor insisted that he must have hit it with a hammer in his work. Dad was equally as insistent that he would have remembered doing that, but said no more.

Both his mother and father died of cancer. His father when he was 11 and his mother a year, almost to the day, before I was born.

He repeated, "I know its cancer. I just know."

My cup fell over and I watched the dark brown contents seep slowly into the grass.

He agreed to go back to the doctors with me on the proviso that I didn't tell anyone else.

More secrets, more acting. That was the easy part.

After numerous consultations, scans and much questioning, Dad was diagnosed with melanoma. The lump was a tumour; the breast is a metastatic site for melanoma.

My own extensive research showed that melanoma often starts with a black line on the thumb or big toe nail.

Following surgery and radiotherapy, Dad came home from hospital. And so did Mum. It had been a grueling time for us all. My brother had continued to run the business as I had taken on the practicalities associated with Dad's illness. I had struggled to understand the consultants and the medical explanations and taken copious notes. Each night I would undertake my own research and try to comprehend the process and probable outcome.

When Dad returned home, he would come and sit in a deckchair wherever we were working; it was nicknamed the Director's Chair. Some days he would be too weak to get out of the van, but at least he was out and about and involved.

Then the universe threw another curved ball.

The girl who had bullied me at school came back home to the village. She too had cancer. I visited her and we reminisced about the past. The bullying was never mentioned. On the days that she felt well enough, I would saddle up two horses, her father would heave her into the saddle and we would go for short rides. Sometimes just the exertion of getting into the saddle was too much. She would slide back into her father's arms and return to the house in her wheelchair and in tears. On those days we would drink coffee, eat cake and watch Black Beauty videos.

I would visit her at Felindre, the Cancer Hospital in Cardiff. It was where Dad had been. Dad was at home now but it was a route I knew well.

After getting permission from the nurses, I would get her the McDonalds meal she craved on my way to the hospital. She would devour it, enjoying every mouthful, and later be incredibly sick. She insisted it was worth it.

Eventually, sitting on the edge of her hospital bed, I plucked up courage to ask her about the bullying. There was never going to be a right time. I clumsily asked her why she had been so horrible to me in school - what had I done to make her hate me so much? I added, of course, that it didn't matter now; I just wanted to know.

She was completely astounded and genuinely had no comprehension of the misery she had caused me.

She thought long and hard whilst picking the gherkins out of her Big Mac. "I know we fell out a bit but you had everything going for you. I suppose I was jealous. I wanted you to be my friend but thought you hated me."

We left it there. She died weeks later aged just 31.

She taught me a lot about perception before she left.

In the meantime, Dad's cancer had returned. It was deemed 'aggressive'. We were once again rebounding back and fore to Felindre as he had chemotherapy. All he was worried about was losing his hair. He never did.

Ironically we were all pleased for him. I have since learnt that hair loss is a good sign; it indicates the cancer cells are responding to the treatment. Bald would have been best.

Mum was hospitalised again. My brother and I kept the business going and would visit Mum in Talgarth one evening and Dad in Felindre the next.

Christmas came. We visited Mum on Christmas morning and then drove 70 miles to Cardiff to visit Dad in the afternoon.

It was a memorable Christmas but for all the wrong reasons.

Around that time I was introduced to a lovely local chap by a good friend. Knowing of my reluctance toward relationships, she had said, "He's lovely, he'll be great company and a good friend. It's what you need now. You don't have to marry him."

But I did.

We had only been dating for a few months when Dad was given a couple of months to live. My new partner had the kindest of hearts and suggested we get married so Dad would at least be able to be at the wedding and know I was settled. He had the best intentions in the world and everyone agreed that, although a little premature, it was a great idea.

It injected a little joy and sunshine into a very dark arena.

One of the hardest times of my life was sitting down one Thursday evening with my Dad, my Mum and my brother to choose hymns for my wedding and Dad's funeral. Dad also chose his bearers and insisted that we didn't mark his passing with a 'bloody bench' or anything like that. He asked, "please don't think of me sat up there on a bloody cloud or anything and I don't want people to visit a tree or a bench with my name on it and cry. Just remember me for who I am."

He sat with his head in his hands, tears rolling down his cheeks and whispered the words that are etched into my memory. "I can't believe I am arranging my own funeral."

I will never, ever forget it.

Dad did make the wedding, walking with two sticks and a full head of hair. The wedding was held at his favourite local Country Hotel, The Gliffaes, where he had also maintained the gardens for years. We flew the few miles in a helicopter, circling above his house and the village first.

He had never been in a helicopter; his was more of a 'tea cup list' than a 'bucket list'.

He insisted that my new husband and I went away for a week to have a break and joked that he would hang on and wait for us to get home before he 'shuffled off'.

I called him every day from France, Portugal, and Spain with details of our motor biking trip, often talking to the answering machine for a while to give him time to get to the phone. And true to his word he hung on.

Two days after we got home, he died.

It was the 14th June; the day before Father's Day.

Death belongs to life as birth does. The walk is in the raising of the foot as in the laying of it down.

\- *Rabindranath Tagore*

36.
Ask and it is given

'Darling you are completely
bonkers but we love you for it.'

As I sat quietly in front of my wood burner with the rain beating on the window, pouring over my Ordinance survey map carpet, I encountered another crashing wave of overwhelm.

I was just a couple of months away from running across a country - without a designated route.

The not-yet-proposed route for my attempt to cross Wales barefoot had become the route of discontent, the route of all evil. As I was now planning to run the opposite way to the route I had been given directions for by the Rotary Club, I had to reverse every instruction.

Downhill became uphill, left became right and 'from tarmac to field' became 'from field onto tarmac'. All of that was proving to be stressful enough but there were numerous gates and stiles involved with many of the directions including instructions like 'go through the gate on the left and turn immediately right up narrow track'. Of course that meant that I would be coming 'down the narrow track, turning left at an unmarked point and going through the gate on the right.'

I was rapidly losing confidence and enthusiasm.

Thankfully other people still had confidence in me and my endeavour. I had approached Harry Legge Bourke, the owner of Glanusk Estate, a local Country Estate and asked if I could train on their quiet web of private Estate roadways.

It had been proving difficult and dangerous running barefoot on the rural, single-tracked country lanes. Glanusk Estate would be a much safer option and was also just down the road.

I had often worked in the formal gardens of the Estate, working for Harry's mother, Sian Legge Bourke and his Grandmother Lady De L'Isle, with my father and my brother. We had been referred to as The A Team. They are a fabulous, accommodating family who do an enormous amount for the community and local area.

In a little thread of irony, it had been Lord Glanusk who opened Mid-Wales Hospital where Mum had spent so much of her time.

Harry Legge Bourke is affable and approachable and when I called with my random request, he guffawed and said, "Darling, you are completely bonkers but we love you for it - yes, of course you can use the estate. Good luck."

As I ran the chipping-strewn roads I would sometimes run through the yard and noticed one of the office units was an inhabited by a company called The Blaythorne Group, their motto was, 'equipping people for life.'

An idea began to form as I wondered if they would support me in my challenge in some way. Perhaps they could 'supply' someone to accompany me on the route and take responsibility for the orienteering and navigation. They specialised in preparing people for hostile environments by providing training and equipment for extreme challenges.

A sign from the universe surely?

There was only a month to go until my challenge weekend and I still had to confirm the route, whether or not I would run alone or with support and even if I was going to attempt to run it in just one day or to take two. Both of the walks across Wales were completed in a single (long) day. It was technically feasible for me to run it in the same amount of time.

There were an awful lot of unknowns. I was going to look slightly unprepared to say the least. But that's why I was approaching them. I need help in the preparation. Move over ego. Come in courage. It was worth a go.

I called in one day after a bare foot run and just poured out my dilemma to a guy behind a desk with a black lab at his feet. I wasn't particularly prepared and I definitely wasn't cool but I was convincing.

Tim, it turned out, had previously completed the Marathon des Sables, a six day ultra-race; the equivalent of six marathons across the Sahara and referred to as the 'toughest foot race on Earth'. He was immediately intrigued with my self-imposed challenge. He promised to put a relevant proposal to the board members and letting me know what they decided.

Not for the first time, I thanked God, the universe and spirits for my ability to just approach people and 'ask' for what I wanted.

37.
Listen and learn

'The most important thing in communication is hearing what isn't being said.'

After Dad's death, my marriage also died. Very quickly. My ex-husband was and still is a kind man and neither of us anticipated the additional emotional complications that followed the upset and upheaval of Dad's passing.

We had married with good intentions but for the wrong reasons. The decoy had suited us both, for different reasons but was unsustainable. I will always be grateful for his support and kindness. Possibly the kindest thing he did was let me keep our little dog, Tippy. And for that I will always be grateful beyond words.

At a complete loss once again, I turned back to work and nature. Nature provided my professional environment and a safe environment for my recreation and recovery.

My ability to communicate with animals evolved rapidly.

Whilst talking to animals is pretty common place and comfortably accepted by most, communicating with animals is a bit different. It is a two-way communication; you listen and hear, or feel, a response. Animals were my friends, my allies and I wasn't arrogant enough to think that it was to be a one-sided relationship. They helped me and I wanted to help them. Why wouldn't I listen?

Everything has a vibration, an energy. We are all connected. Of course my preference was to connect to animals but I soon discovered that it was possible to create that same energy exchange or communication with plants and even rocks. Later I was to learn that the same vibratory connection was possible with computers and machinery.

It was to be particularly useful. More than once in work, I have explained to a mower which refuses to start, that it is obviously it's prerogative whether or not it chooses to work however, if it doesn't then I will have to replace it with one that is more willing and reliable. The boys in work have witnessed a mower spark into life after this little pep talk.

Try it with your computer. When a computer or printer plays up we immediately get frustrated and direct a negative surge of energy, expletives, disappointment, frustration and hatred toward the machine. This just increases the negative energy around it.

Try being nice.

Explain you understand it may not be having a good day but you need it (we all want to be wanted), it is important to you; you appreciate it and would be really grateful if it could assist you in your work. Be light, have fun, uplift the energy and be prepared to be amazed at the results. Adopting this attitude will be of far more benefit to you and the machine.

When I explain these unconventional principles that I employ in my life, it evokes one of three reactions:

There are those who are not surprised as they communicate in the same way. They are thrilled to find a like-mind and we enjoy interesting and inspiring conversations.

Then there are those who dismiss the notion totally. They are closed off to any possibilities outside of their reality.

And then they are those who are dubious, skeptical but also receptive and intrigued to know more. The third group will ask questions. My favourite question to be asked is, "How does it work?"

I always smile and reply, "Very well, thank you."

Others ask, '"How do you do it?"

Like a lot of things, I find doing is easier than explaining. Helen Keller said, "The best and most beautiful things in the world cannot be seen or even touched. They must be felt with the heart."

I am becoming far less inclined to either look for, or offer, an explanation, preferring to simply share on my experiences and relevant information with others for them to develop or discard as they wish.

"When the pupil is ready, the teacher will appear."

Most pet owners already talk to, or at, their pets.

Walkies? Dinner? Are the most obvious examples but sadly they underestimate the pleasure to be gained from listening for a response and conversing.

When I encourage them to do so, I am frequently told, "I think I get a response but I think it's my own mind, my own thoughts."

Why would it be? Trust what you hear or feel. It is the hardest part but also the most rewarding.

For example, when I ask my Westie Yogi if she 'fancies a walk?' I will get responses that include, 'Yeah! Let's go." Or maybe a sarcastic, "What do you think?" or a more civilised, "Can we wait until it stops raining?"

All of these responses are easily attributed to being own thoughts but I don't believe they are. The more you develop a conversation the more you will discover the real personality of your animal. It is a beautiful process and they are always so grateful. Well, usually grateful. I have had some cats be very aloof and prefer not to talk with humans as they think we are rather stupid and I have had some depressed animals which can be upsetting, just as a depressed human can be.

Never underestimate what you will hear. I have had a dog describe his owner having an affair and other numerous difficult situations have been revealed. Remember your pet will pick up on everything that happens in the home. Sadly there are many pets that have to keep secrets, which compromise their own happiness hugely.

The wonderful part for me is that the more you practice, the wider range of answers and even attitudes you are able to detect, receive and trust, which can be comical. The quicker you can engage with this element of fun, the easier it will all become. Maintain an air of lightness and 'ease', animals and plants are generally very uncomplicated in their communication.

It doesn't mean you can't have serious conversations - animals and plants also have much to teach us - but they tend to be far more direct and straight talking than we are. I find this refreshing and often comical in contrast to our well trained and ingrained polite and politically correct tendencies One of the reasons you may doubt what you hear when you start is that answers may be frivolous and even rude. It's as though our plants and animals and even rocks and inanimate objects don't have a filtering process - they say it as it is.

Obviously I chat with my own animals in the same uninhibited manner as I do humans and never cease to be amazed and impressed by the honesty and uncomplicated attitude of the former. It's not that they lack emotion - far from it, they just lack that filter.

A lot of dogs I speak to are suffering from depression to a degree, they want to be more involved in the family and want to be considered far more than they are. One of the things I get most from the animals is, "Why don't they just ask me, it's as though I don't exist."

I try to explain our fear. Fear of getting things wrong, fear of looking foolish, fear of being judged and so on.

Animals, like young children, naturally live in the moment. They don't understand our preoccupation with the future and in particular, worry. Try explaining about paying bills and mortgage and they will still bring it back to the basic, "so how does worry pay the bills?"

Most of the time, I run out of answers before the animal runs out of questions. You simply cannot disregard their rational reasoning. It is basic and simple yet authentic and congruent.

They can teach us so much.

Whilst I find that animals love the stimulation of day to day chat, plants, trees and rocks tend not to want to be bothered by such trivial things. To me they are all the great guides of this World; they are prophetic and wise and organically want to share that wisdom.

As with animals, all you have to do is ask and listen.

Again it is the understanding and trusting what you hear that can be the hardest part. Believe me - you will not always hear what you want to or expect to.

For me, communing with rocks was the most unexpected intuitive development and I was thrilled to be guided toward it by a lovely book, Conversations with Rocks by American writer, Ariana Houle. It was and still is of great reassurance to me as I journey along this path. She explains that almost all of us will have at some time picked up a special rock or stone. In actual fact we haven't chosen it, it has chosen us and as it sits quietly in our pocket or on our desk or drawer, it will be imparting its own special guidance for us. Of course once we know this we can commune with it if we choose. And often our special 'rock guide' will be one that we go and sit with on a favourite walk.

And of course there are the plant spirits and elementals. Oh boy, they are great to converse with.

For years I have communed with the nature spirits of a garden prior to working in it. It is amazing what information they can give you about the immediate area.

I also ask their permission to carry out work and have numerous examples of both positive and negative responses. Some of which have been experienced by others on site too.

I have met many people who journey with plants and animals in the traditional Shamanic way. Following the beat of a drum you can meet with a spirit animal or guide and ask for guidance or help with a certain issue. I too have done it often and it is indeed powerful but it is different to the relationship I have with my elemental allies.

I communicate with animals, rocks, spirits and indeed whatever wants to be heard, in a very unstructured and organic way.

They are always there. I don't have to journey or drum them up.

It is as though we all walk together, peacefully, until someone has something to contribute to the silence.

People often ask, "What messages do you get? Do they have a message for me?" And while I am fortunate to feel guided and protected, I do not use them as a daily horoscope. We communicate as you would with human friends, sometimes discussing every day events, sometimes sitting in silence.

You wouldn't keep asking a friend, "Do you have a message for me?" I don't keep asking my spirit friends either.

Some people are disappointed when I describe something they perceive as being extraordinary, as being so ordinary to me. I believe these kinds of connection are there for all of us to rekindle. We were born with the ability, some chose to embrace it; others turn their back on it.

The stark reality is of course that I am incredibly confident of this aspect of my life and of sharing it whilst equally aware that if my mother was to share similar beliefs she would risk assessment by various mental health staff. Ever aware and cautious I put this conundrum to a GP. He replied, "If it were your mother talking in this way, it would be deemed out of character. Whereas it is your character, it is your very essence."

I recall having a strong connection with animals and spirits as a child, before Mum was diagnosed with bi-polar. I also recall consciously deciding to keep quiet about it.

38.
Intuition - tuition from within

'My head knew it was a sensible option.
My heart sank at the thought of taking it.'

It was the mighty Hannibal who said, "We will either find a way, or make one." Unfortunately he wasn't available to assist in my plans to find my way across Wales. Busy with some elephants or something.

However, following my unstructured, spontaneous plea to Tim at The Blaythorne Group, I was invited to meet with the board members and deliver my proposal; it was like standing in the Dragon's Den.

One of the board members, Nigel, had run the Walk Across Wales Route the previous year with a Welsh TV presenter, for charity. He had assisted in her training and also supported her through the run as a running buddy.

He confirmed it was indeed a substantial challenge and they had both, obviously, worn trainers.

Various possibilities were mentioned from providing drones to follow, monitor and record my progress to finding a running buddy to accompany me and who would take responsibility for navigating.

It was also suggested that I complete the run in one go, without breaking overnight. In principle it was doable and concern was raised that if I stop for any significant length of time, my feet may swell and put pay to any running the following day. However, doing it in a single stretch may mean tackling some of the route in the dark. Not an attractive thought without footwear.

Far from being reassured by the meeting, I left feeling even more overwhelmed. I was reeling under the enormity of it all. Ironically, I now felt more pressure to conform and perform. There were even more details and practicalities to address and consider.

However, some things also became crystal clear. I didn't actually want to run with anyone. I wanted to cross the country on my own. That was the whole point. To indulge myself in the solitude, facing my fears and conquering emotional and practical obstacles, quietly and alone.

I also realised it was totally unrealistic at this stage.

Nigel gave me his business card and said he would help as much as he could. He was unable to run with me as he was training for his next self-imposed ultra-challenge.

He proved to be an inspirational ally and mighty motivator. When we met just a week after the board meeting I was at very low ebb; my days were completely consumed with running, planning and working. I was already tired and still had an awful lot to sort out and confirm. I couldn't even decide what I wanted to do, let alone how I was going to do it.

Did I run with a buddy or try to find my way across the country alone?

Another option had been raised. I could join the organised Rotary Walk, as I would have the benefits of the events support and organisation.

There would be many advantages to joining this organised walk, from a cooked breakfast (honestly), route markers, company, support and hopefully unnecessary First Aiders!

But would I benefit from the camaraderie and structure of the organized walk? I am a solo player. It wasn't what I wanted. It was moving further and further away from my original motivation.

However, the organized walk was being held on the Summer Solstice. That appealed to me greatly. What a beautiful time to complete such a monumental challenge.

Running the walk wouldn't be a problem. I ran at the same pace as a fast walk. And there were also participants who ran the course but I also had to incorporate those extra 6 miles to the coast and I would go against my will, running inland away from the sea.

My head knew it was a sensible option. My heart sank at the thought of taking it.

Nigel was calm, pragmatic and congruent. We chatted and I confessed to finding the whole journey achingly lonely and yet still preferring to undertake the challenge in solitude rather than with company and close support.

He understood it all it totally. He encouraged me to listen to my heart rather than my head. It would be more reliable.

I suspect he has no idea just how instrumental he was in the planning, and to the success, of the whole challenge.

39.
For better; for worse

'"Hiya," she beamed, "meet my new husband. We've just got married."'

1996 was a 'solid' year. Solid as in hard. Hard as in teak. The numerological value equates to seven: "It is the essence of the seven that most symbolizes our struggle to know and understand."

Amid the mind-wringing agony and aftermath that followed my Dad's death and my divorce, Mum was sectioned again.

It had been a brutal time for us all and had also triggered another manic stage of Mum's bi-polar.

Mid-Wales Hospital, or Talgarth, had closed and Mum was now being admitted to the Acute Mental Health ward at Bronllys Hospital. It was in the same area geographically but miles apart in conditions. For the better.

A friend had come to stay with me for a few days and while we were catching up on each other's news late on a Saturday morning, there was a knock on my door.

My jaw dropped to see Mum standing on the doorstep with a man by her side. I vaguely recognised him as a patient from Bronllys.

Mum was uncharacteristically dressed up, wearing a skirt and make up. He was in a suit.

They were holding hands.

"Hiya," she beamed, "meet my new husband. We've just got married."

Dad had died less than three months ago. She had been in hospital for most of that time, under section for half of it. And yet she and her new husband, a fellow patient being treated for schizophrenia, had managed to arrange their marriage at Brecon Registry Office and go through with it without any of the medical staff at the hospital , or family, knowing.

In the chaos that ensued, the marriage was eventually annulled. She still lost her widow's pension which Dad had worked hard for. We were advised to take legal advice but didn't have the energy, the knowledge or incentive to even know where to begin. There had been numerous dubiously-handled incidents and many times my brother and I had sat listening to Mum's psychiatrist wondering if he was a patient who had found a white coat.

During my hot-headed moments of wanting justice, Dad had often quietly reminded me that we would need the Mental Health system, the hospital and the staff in the future.

We said nothing.

The pain and confusion was becoming impossible to deal with and work failed to numb it any longer. I had to find another distraction, another cause to focus on to avoid my own turmoil.

I started fund-raising for cancer charities.

Not overly keen on joining organised events, like marathons, I started creating and organising my own. I arranged and organised a road trip to Iceland to run the Reykjavik Marathon. Raising money was only part of it of course; I didn't want to spend any either. That would be counter-productive.

I set about getting every aspect of it sponsored. I didn't know if it was possible but it was worth a go. I was used to asking for things.

Transport was a priority and as a result of a polite phone call, 4x4 retailers, M. Burgin and Son, in mid-Wales loaned me a brand new Nissan Nivara double cab to drive.

I am even more astounded by people's generosity as I look back on these events than I was at the time.

I would simply cold-call people, explaining my intention and how I thought they could contribute. So, would they, please? Most would and did. The Burgins also covered the truck in decals for me and gave me a morning's off-road tuition. It was a great advert for us all.

Smyril Line generously agreed to sponsor me by gifting the ferry crossings. A little bored on board, I asked if I could possibly go up on the bridge and meet the Captain. I have no idea where these ideas came from, let alone the courage to ask. But came they did and once again I was warmly welcomed. I enjoyed a couple of hours in the most comfortable black leather chair up on the bridge, learning the significance of the various dials and displays in front of me.

The dashboard even displayed the temperature of the fryers in the kitchen. I was mostly intrigued with the more practical aspects, like reversing, which was still done on huge wing mirrors. The Captain seemed happy to keep answering my enthusiastic questions. I kept asking.

Once in Iceland, I found my way to Reykjavik, ran the marathon and then drove on down to the Blue Lagoon. Not surprisingly, my ability to ask produced many opportunities. As I had driven a large part of the circumference of the country I decided it would be a good idea to see if I could do it all. I did. Often negotiating dirt roads and even no roads, by following rather bemused locals.

I met with Iceland's major motoring celebrity, the equivalent of Jeremy Clarkson, and was invited to join in various off road manic sports. I was interviewed for Icelandic news and my sign written truck, 'from Wales to Iceland' was recognised wherever I went.

I decided that if Wales was God's Pocket, then Iceland was God's Playground.

All in all I was treated exceptionally well as I quietly navigated my way around the stunning countryside, venturing up in to frozen volcanic craters and negotiating sheets of ice in my trusty 4x4. I felt divinely protected and had a ball.

I also raised over five thousand pounds.

With time always at a premium, I needed to create a few less time-consuming fundraisers too.

Locals at the Horse Shoe Inn, where I was working behind the bar, suggested I compile and feature in a calendar as a fundraiser which would require less physical effort than most of my previous challenges.

They suggested I posed naked covering my bits and bobs with flowerpots, gardening tools and my dog. It was 2002 and before the WI-esque Calendar Girls film was made, so still original and innovative.

A friend took some photos of me one weekend. We used my back bedroom and whatever we could find as props to keep it as low budget as possible.

We called it 'Bare Rooted.'

It was just one letter different to 'Bare Footed'. Another sign?

A local printer created and printed the calendar for me and I set about selling the 100 copies to the locals in the pub and friends.

It was pre my internet days, so sales were slow and steady. If I sold them all I would raise £500 for Charity; I had chosen the Arc Addington Fund. It was at the time when Foot and Mouth was devastating the countryside and I wanted to raise awareness and funds. As a result 'Bare Rooted' was featured and promoted in most of the nationwide Farming and Countryside magazines. Consequently, I was invited to the Winter Fair at Builth Wells by the National Farmers Union to brighten up their stand. "You could sell your calendars there too," they added.

It was a bitterly cold day and the showground is one of the draughtiest of places on a good day. I was nursing the remnants of a cold and not feeling great. I had only five calendars left to sell and was looking forward to being able to go home, when I noticed a commotion in the aisle.

A knot of people were approaching. Security men cleared a path for them and there were cameras, reporters and a general kerfuffle.

Someone behind me whispered, "It's Prince Charles. I dare you to give him a calendar."

Without having time to think it through, I stepped directly into his path and offered him one of my calendars.

He was delightful. He had the twinkliest blue eyes, was utterly charming and had a wonderful sense of humour.

He flicked through the pages and said, "My word, that's remarkable. What a wonderful way to raise money, by getting One's kit orf."

He added that he would probably pass it on to William.

As he walked on still clutching his calendar, I was immediately swallowed up in a wave of people and cameras. They also bought the remaining calendars. Result.

The following day the newspapers contained photos from the calendar with a variety of headlines. The Telegraph read, "The Prince and the Gardener"; the Daily Mail read, "Calendar Girl catches Prince Charles's Eye" and The Sun read, "Prince Charles tells Lynne to get her kit orf."

It was a surreal few seconds that changed my life radically.

40.
No mean feat

'TV presenter, Matt Johnson, was very supportive and rather impressed. He gave me a big hug and trod on my left foot.'

It was mid-April, just six weeks before my barefoot challenge was due to take place, when I made my first appearance running bare foot at a local public event.

It was a 10K run for Sport Relief at Cardiff Bay, with approximately two thousand shod runners and with a mile long circuit through the city being run six times. BBC Radio Wales asked me to not only record my experience but also to interview other participants and supporting celebrities.

BBC Radio Wales had been remarkably supportive of my challenge and had been following my progress on Eleri Sion's afternoon programme on a regular basis. I was glad to take on the role of interviewer instead of interviewee for a change as it gave me something, other than my vulnerable, bare naked feet, to think about.

Holding a rather important looking microphone also made me look a little more professional and a lot less mad.

I found myself far more nervous than I expected, so was increasingly grateful of having a job to do.

I decided to turn up barefoot rather than shed my shoes on site. That decision led to my first incident. I walked quickly and quietly into the foyer of the Millennium Centre en route to the media room. I didn't want to draw any attention just yet, choosing to just find my feet first.

A couple of steps through the heavy glass doors, I realized that the bottoms of my feet were wet from the rain soaked pavements. In exactly the same second, I realized the flooring was of polished marble. If you have ever stepped out of a bath onto a marble floor in a posh hotel bathroom you will know exactly what happened next. I was like Bambi on ice. I managed to stay upright but only amid a lot of arm flailing, hip twisting and high knees.

There is nothing quite like an understated entrance and that was nothing like one.

Despite having been on Welsh Radio regularly and having already received a reasonable amount of 'exposure' about my exposed feet, I still felt unexpectedly 'abnormal'. People stared. And pointed. A lot.

Acting my way out of my un-comfort zone, I managed to corner a few celebrities in the safety of the carpeted media room. A deep breath and big smile belied my emotional discomfort. I held out my FlashMic like an olive branch.

TV presenter, Matt Johnson, was very supportive and rather impressed. He gave me a big hug and trod on my left foot. Jason Mohammed still thought I was completely bonkers and the 800m Olympian runner, Gareth Warburton, was adamant that he would never be tempted to try it. He wasn't the only one. Not one person expressed even a flicker of interest in trying it for themselves.

Once down at the start, the general public seemed to have the same views: that I was bonkers and they had absolutely no desire or inclination to try it!

Humour had to be deployed. Armed with my FlashMic and a big grin, I asked a few fellow runners whether they thought I was bonkers or brave. Bonkers was the general consensus of opinion. One little lad very wisely added, "You have to be brave to be bonkers."

I laughed. It was nervous laughter. Would I actually be able to do this? I had entered for the full 10K although was obviously free to stop at any time. In theory.

I had no idea what the route surface was like. I had tried to Google the route but it gave no detail of the surface for anyone running bare foot - funny that. I had deduced that at least half of it was tarmac and just hoped it was fairly new and therefore smooth. It was a city, surely it would be.

It wasn't, it was awful. The type I had nicknamed 'Terror Tarmac'. It was open, old, gritty, sharp and pitted. The very worst type of tarmac. The type that shreds your feet very quickly, especially wet feet. White lines (nice and smooth to run along barefoot) were of a premium and amongst a jostling group of 2,000 (shod) runners they were not easy to access and utilise. As they were at the side and middle of the road, either side of tyre pathways, they were also glistening with shards of broken glass.

The hostile tarmac led into a stretch of anti-slip block pavers. Anti-slip is normally good, if you are shod, but in order to be classed anti-slip they are quite severely scuffed to give a rough surface for maximum purchase; not so good for bare skin.

That in turn morphed into smooth block pavers which, after another shower of rain, became my second skating experience of the morning. The final stretch was an excruciatingly painful length of wooden grooved decking boards dusted with a light sprinkling of anti-slip gravel. Good grief, where was the red carpet? Or any colour carpet for that matter?

I'd like to say the first circuit was the worst, but of course it wasn't. It had merely served as a recce and a sobering one at that. Now I knew what to expect for another five laps. With one hostile section following another, I tried to take my mind off my discomfort by answering the questions from fellow runners. And also asking them.

I was encouraged by a few friendly faces which had turned out to support me and it was lovely to have a quick chat with RWA director, Tom Wood, and his daughter, Evie. After the event Tom told me that Evie had said she wanted to start running so she could do next year's Sport Relief Mile. That is inspiration at its swiftest and most rewarding.

Other runners were happy to talk as I waved the mic in their face whilst trying to pick my way precariously through chippings, broken glass and dodging all the trainers thumping around my little bare tootsies. It seemed everyone was questioning, and testing, my sanity. Although with each lap I completed, the stewards slowly transitioned from shaking their heads and wincing as I passed, to clapping and cheering.

I had expected to run the 10K in about an hour and a half but actually managed just over an hour, so was pleasantly surprised. My feet were fine. No cuts, blisters or sore points.

It all bode well.

41.
TV? Me?

'Three months later I was standing in front of a camera presenting a garden make-over series for BBC Wales.'

After my encounter with Prince Charles and still reeling from all the attention and leg-pulling, I got home from work, cold and wet, a few days later to find an unexpected message on the answer phone.

"Hi, I'm calling from BBC Wales. I wondered if you were interested in discussing a TV programme idea we have for you."

It was one of many messages referring to the calendar. Most were from friends who were teasing or trying to order a copy for themselves.

I was cold and tired. I thought it was another leg-pull or well-meaning wind up. I ignored it. Another message was left by the same researcher a few days later.

I called them back apologetically.

Three months later I was standing in front of a camera presenting a garden make-over series for BBC Wales. The lads I worked with on a daily basis in my business had agreed to be the make-over team and that, of course, included my brother.

I had attended several meetings with the BBC and, as a result, the programme format had been created, commissioned and titled 'Hot Plots'. With a researcher, I found a suitable garden, then created a design, gave the director a list of materials we would need, scheduled in the boys to do the work and then just did what I loved to do, in front of a camera. It was a dream. We were all being filmed doing what we loved and what we did on a daily basis without a crew watching us - camaraderie, leg pulling and all.

And there was another strange twist, one of those delicious synchronistic signs that you just cannot ignore.

The cameraman who filmed 'Hot Plots' had filmed a documentary on volcanoes a few years previously.

He started to retell a story about a tiny little island which none of us would have heard of called Montserrat. He had spent a week there filming the aftermath of their volcanic eruption and had met many of the people that I had lived amongst. He was able to tell me how they were and what had happened to them and the island.

'Hot Plots' ran for 12 programmes and was a lot of fun. When it was broadcast I got a call from a York-based Production Company and went on to film a series of garden make-over programmes called 'Gardening Angels'. As well as co-presenting with Mark Curry of Blue Peter fame, I was also responsible for providing designs for the make-overs from emailed photographs. We filmed fifteen programmes in eighteen days. It was grueling. I loved the work, I loved the challenge but I didn't like the production team or crew. I was considered an outsider. One of the main landscapers in York had been demoted from designer to make way for me. He wasn't happy and didn't disguise it. It was like being back in school. I felt that fear and dread rising up again. I stuck it out, politely refusing to be involved in a second series.

"When one door closes, another one opens. But the sensible thing to do is to get as far as you can whilst you still have some light from that first door."

In light of my TV work, I was asked by the Western Mail, Wales's National newspaper, if I would contribute a small weekly column for them. Of course I would. I would be delighted to. We agreed it would be called 'Gadget Girl'. I wrote about all the laddish gadgets I loved and got to use in my varied outdoor work.

The column would go on to grow organically into a full page weekly feature called 'Green Scene'. I always have been and remain grateful for the opportunity. I have never studied writing, just loved it. I love sharing the wonderful things I stumble across. Yet I still don't consider myself to be a professional writer, just a passionate one.

I also loved adding columnist to my ever-growing list of work. When anyone asked what I did, I would cheerfully reply, "Which day are you interested in?"

As well as creating challenges for charity, I continued to keep up a personal stream of achievements, too. I took and passed my motorbike test with my brother. We opted for the four day intensive course just one week before the legislation changed to make it a more prolonged training process. We both took our test with snow on the ground and both passed first time.

I bought a little VFR NC30 400cc and he bought a CBR 600cc, so now not only did we work together but we rode out on our bikes together on the weekends. It was great. I am very close to my brother, respecting him and admiring him hugely.

I loved riding a motorbike; the solitude of it, yet being even more connected to everything around you, as your sense of awareness is heightened. Everyone should learn to ride on two wheels before being able to drive with four. The roads would be safer as you learn self-preservation, how to read the road, anticipate the actions of other road users and not to rely on the tin box around you to keep you safe.

I loved my life. Every day was different. One day working on site with a load of lads and then the next going into the studio to speak on a radio show, or speaking for a group of WI women at a lunch, or even just sitting quietly at home researching and compiling my weekly columns.

I loved it all.

Following the success of my Bare Rooted calendar, I was approached by a local newspaper to compile another to promote an anniversary. It was fun and a fabulous smoke screen for my always ailing self-confidence. Acting 'as if' was easy.

A third calendar, raising funds for Marie Curie followed; another sell out and well supported this time by the Welsh rugby team. Happy times indeed but I decided it was time to hang up my calendar career. After all, I had completed the hat-trick.

I was also being recognised and approached in public, often being asked for my autograph. It was a crazy scenario for a little girl from a small village, who had the most crippling self-confidence issues. I never managed to write an autograph without asking, "Are you sure?" first.

A modelling agency in Cardiff offered to put me on their books. They would send a photographer to take a few suitable shots, put me on the books and let me know if work turned up.

Following the TV success I still felt embarrassed when people remarked, "I saw you on TV last night."

I had to think of something funny to reply; a retort to disguise my wonderment of it all. My brother did it for me. "On Crime Watch?" he would tease.

It became my stock answer too, so when I got a telephone call from BBC TV in Birmingham saying that they produced Crime Watch and would I please be a stand in for a reconstruction they were filming the next week, I laughed and put the phone down.

They called back.

They thought we had been cut off and repeated their request.

"Ha ha, great gag," I retorted. "I appreciate it!" And hung up again.

I was convinced it was one of the boys in work taking leg-pulling up a level. The researcher patiently dialed back a third time a little confused. "Please call me back," he said, giving me a number. I did and got straight through to the receptionist and transferred to the bemused researcher. It had been a genuine request.

I apologised profusely - of course I would do it.

They had obtained my details from the agency in Cardiff; I loosely fitted the description of a girl seen standing on a street corner whilst a young man had been savagely attacked under a railway bridge in Cardiff. She was of slight build, had long blonde hair and had been wearing a long red leather coat. It was presumed she was a prostitute.

I would be on national TV playing the part of a prostitute in a reconstruction.

Another rather surreal installment in my life and the fuel for an awful lot more good-humoured leg-pulling.

We just don't know what is around the corner - or on it.

42.
Gravel, grit and someone quits

'The shoes had been off for a while
but now the gloves where off too.'

As interest increased during the last few weeks of preparation for my barefoot run across Wales, I was being asked constantly about how the training was going. The honest answer sounded ridiculously evasive: "I don't know; I have nothing to measure it by."

I was finding it very difficult to assign the time needed to get long runs done. I was comfortable doing eight miles on tarmac but it wasn't far enough. I was comfortable doing fifteen mile runs in Vibrams but that was not far enough either.

I only had six weeks to go. Or just 42 days to train.

The irony was that as the weather improved and daylight increased, so did my workload. The weather seemed intent on tripping me up too; rain, strong winds, fog and frosts had all made my favourite mountain road running route out-of-bounds. Running lower roads meant more traffic and that was not a good mix with barefoot running!

I had completed several mountain runs which were exhilarating and encouraging until I realised that my longest single run barefoot had been just ten miles. It felt so much longer - twice that amount - probably due to the two separate ascents of 2,000 ft. It was an explanation, not an excuse.

On the 12th April I wrote in my diary: "Sometimes I get a training run just right. I eat the right thing at the right time beforehand, I'm not tired starting, I have drunk enough, the sun is shining, the roads are dry and it all goes like a delicious dream. I feel like Forest Gump. I could run forever.

Once in my stride I feel lighter and stronger than ever and it is so exhilarating to recognise that I am connecting with the earth energy with every bare step I take. It's pure magic. I remember to keep my shoulders back and relaxed, lift my knees, pull my rear foot up, keep my head high and look straight ahead. I remember and follow Mike's three key points - RRP - relax, rhythm and posture. It all just falls into place. And that is what keeps me running barefoot.

However I guess the fact that it is called a challenge suggests it is not always as smooth as that. Largely, the training is proving to be an uphill slog - psychologically as well as practically."

That same week, I had been determined to see exactly what my feet were capable of. I had been erring on the side of caution as I didn't want to have to devote any valuable training time to recovery or healing time. But now I had to know where I was at. I might be pleasantly surprised. Or not.

I chose my tried and tested tarmac route over the mountain. Not only did that limit the amount of traffic I would encounter but I had a definite mile marked out; I could repeat running that stretch to be able to gauge my exact running limit.

I finished work and popped home to change into my running gear. I was already tired. I chided myself back into the truck, drove up to the mountain, kicked off my boots and started running barefoot at 6pm. Three hours and nearly fifteen miles later it was dark, cold and my feet hurt.

I drove home. Intrepid investigation showed a couple of blisters on the ball and toes of my feet. At least I hadn't been heel striking; my heels were intact, but it was still a blow. The first blisters of a four month training period. The first blisters I had encountered in my whole barefoot training period.

I had intended to do the same training the following day just to see how my feet recovered and responded to two days running. They had other ideas and they had already let me know! There was no way that training the following day would be beneficial to anyone other than a foot doctor.

Nonetheless, I was on a mission to fact-find. Suddenly I felt way more serious and determined to face this whole thing feet first. Realising that my tarmac running limit in bare feet was only fifteen miles had spurred me on to another level. I could wallow and worry or I could get into the ring and fight the fears that had been silently swelling inside.

Tomorrow I would drive back up to mid-Wales and recce another part of the route. The shoes had been off for a while but now the gloves were off too.

My throbbing feet had kept me awake most of the night and even depressing the clutch was painful but I got to mid-Wales in the early hours of the following morning. I had driven through a thick blanket of fog and low cloud. It was drizzling and cold.

Ye Gods, why hadn't I said I would run across a Caribbean island?

After wrestling with the map and my enthusiasm, I recce'd several parts of the route which I could access with the truck. My head would not co-operate. I kept getting lost.

Maybe it would be easier on foot. I parked up and ventured into a proposed forestry section.

And sank into a deep depression.

It was a horrifically hostile terrain. The very worst. The hardcore track was loose; it would be like running over Lego. Brambles weaved barriers across the track at various points.

Encountering and including this sort of terrain meant that barefoot shoes of some sort would have to be deployed at these points.

I had seen enough and turned for home. It had not been a pleasant day. Worthwhile, but not pleasant.

Back home, I sat and re-evaluated what I had learned that weekend.

My feet were only ready for twelve tarmac miles.

The route, which was still totally undecided, was unlikely to allow bare feet access all the way.

It was a double blow. I went to bed and the pain in my feet kept me awake for a second night. Little did I know worse was to come.

To quote P.G. Wodehouse: "Unseen in the background, Fate was quietly slipping lead into the boxing glove."

Pertinent on many levels.

I received a phone call.

Gary, the boxer who had helped me get through the pain and grief following Sally's suicide 30 years ago, had passed away.

He was 54 years old.

He had taken his own life.

He had quit.

The grief I felt was overwhelming and caught me completely unawares. I was immediately transported back to all sorts of harboured pain and tortuous memories. Waves of guilt, regret, anger, helplessness, bewilderment and just raw, jagged pain washed through my days.

I stepped out of the ring, sat on the benches and buried my head in my hands. It hurt. It hurt far too much. Did we ever get a Certificate of Competence for life, or just one of attendance?

I questioned him. I cursed him. I empathised with him and then I did a deal with him. He would run with me across Wales. Damn him and damn him quitting. He could bloody well run with me. In spirit.

And so could Sally for that matter. And Dad and every other spirit I could call upon.

Energised by my new team, I stepped back into the ring and upped my game. Bring it on.

I spent the following week discussing various barefoot shoes versus bare foot options with a number of people whose opinion I respect. The overwhelming consensus of opinion was that it was perfectly acceptable to use barefoot shoes for inhospitable and dangerous parts of the route. I researched a number of other barefoot challenges that had been completed by other runners. They all involved the responsible use of a minimal shoe in part.

After all, barefoot shoes provide no support, no cushioning, no padding and require a totally different running style. They also provide protection against friction and puncture wounds.

After all, surely all challenges include the challenge to make any self-imposed challenge as attainable as possible. That's part of… er, the challenge.

That was OK then.

I was ticking the boxes.

I was now running injury-free and staying injury free. Tick.

I had only started running barefoot at the beginning of December, so was running injury-free in just four months. The most common cause of injury in any style of running, but especially barefoot running, is something known as TMTS, Too Much Too Soon. All the minor injuries I had sustained were all down to TMTS and thankfully I had managed to overcome them quickly by various methods and no shortage of will power.

For all the boxes that could be ticked, there was still one outstanding.

An outstanding dilemma.

Just one small thing.

The Route.

Where was it going and how was I going to follow it.

Clean out of ideas, I prayed.

43.
From the dark into the limelight

'I was nominated for a Woman of the Year award and invited to a rather salubrious lunch in London.'

I stumbled into public figure status in my thirties and was always grateful it hadn't been before. I'm not sure I was equipped for it anyway but had it arrived earlier I shudder to think of the mess I may have made of it. I was now being asked to open various events, to speak at various dinners and clubs and join panel shows and debates. I had evolved into a chameleon-esque type character, changing clothes and mindsets to suit the occasion. I would finish a hard day out on site, dash home, feed the animals, jump in the bath, put on a long evening dress, put my hair in curlers, wrap my dress around my waist and step back into my muddy steel toe capped boots to negotiate the little lane to my truck with my make-up and stilettos in hand.

I would then drive to whatever charity ball or function I was due to attend, pulling over in a lay-by before the event to put on my make up and take my curlers out. Arriving at the venue, I would park in a dark corner of the car park so I could swap my work boots for stilettos, unwrapping and flattening my dress as I tottered into another world.

Upon honouring my commitment, the stilettos would be thrown unceremoniously in the back of the truck, work boots donned and I would drive home trying to work out what tools and machinery I needed to load the van with within just a few hours. Occasionally I didn't even make it to bed but would just crash out for an hour or two on the sofa before changing out of a sparkly dress back into my work clothes.

It was exhausting and exciting in equal measures and, not for the first time, was fuelling my love of duality. Two lives. I enjoyed the stark contrasts between my worlds.

Once I arrived at a hotel venue to find I had picked up two totally different shoes. My work boots were far too muddy to use. I opted for bare feet. On stage and a little relaxed by my grounded status I decided to ad lib a bit and shared my footwear faux pas by lifting the hemline of my glittery dress to reveal my Sandie Shaw status. After an agonising couple of seconds of silence, it went down a storm.

I learned that truth is often funnier than fiction and a lot easier that trying to find a suitable well-rehearsed introduction. I was also beginning to glimpse the beauty and necessity of being congruent and authentic. It was years prior to stepping into any bare footing endeavours, but nonetheless, the relaxed and authentic outcome of that evening must have been a small sign sent by the universe. And one which I obviously missed.

Despite missing a couple of universal signposts, I was en route to becoming more authentic and happier in my own skin. Later that year, I was voted the Sexiest Outdoor Worker in Wales in a public poll held by the Western Mail. It was another reminder that authenticity had its rewards.

And there was more.

I was nominated for a Woman of the Year award and invited to a rather salubrious lunch in London. I chatted to Maureen Lipton and Esther Rantzen about growing veggies and other gardening topics.

Gardening proved to be a great leveller. It is something that everyone has some level of involvement with, even if it was just memories of a grandparents green fingers, a passion for it themselves, or even a dislike for it. But the trump card is that almost everyone would like to be able to grow stuff, and being able to often gave me an unexpected audience.

I became intrigued by people's preferences of particular plants and soon realised that their plant choice could often be correctly predicted by their personality.

So I wrote a book about it.

Personali-trees offered a tongue-in-cheek description of people's personality based on their favourite plant or tree. I found the accuracy fascinating.

And so did the Western Mail, for which I was still writing a regular column. They featured it as a double page spread in their Saturday edition.

That very evening I had a call from a man who asked if I would be interested in getting involved with quite a large project he was involved with, in Wales. "It would be a good springboard for your career," he added.

I told him politely that I already had all the springboards I could jump from, thanked him and put the phone down. He rang again later that evening and asked if we could meet in his office during the following week to discuss the possibility a bit further. I explained that my week was already planned out but he could come to my 'office' on the weekend. It would be a mountain.

I always tried to indulge in a day-long mountain walk on the weekend with my best friend - my dog, Tippy. It was a great way to clear my head of redundant thoughts and refill it with more productive stuff which might be useful for the following week.

He agreed and we had our first meeting on a beautiful, crisp day, under blue skies, high on the spectacular Carmarthen Fans.

Unbeknown to me, when he had seen my photo in the Western Mail article, he had told his best friend, "That is the girl I am going to marry."

His best friend had tried to dissuade him. Had he succeeded, he would quite possibly have become my best friend too.

He didn't.

I received the proposal.

I said yes. Well, he had asked.

I should have said no. But he had asked. I felt obligated. Ridiculous I know - now.

There is a saying: "marry in haste repent at leisure." My own experience is: "marry in haste, repent pretty bloody quickly too."

Once again I had been swept off my feet only to be sweeping up the pieces of another shattered episode all too soon. Inadvertently I had climbed back aboard the posh bus: lunches at The House of Commons, dinners at the Athenaeum Club in Pall Mall, more brandy over coffee, more silver. Once again I was a novelty. A blonde Welsh girl dressed in a Chanel suit that could advise Betty Boothroyd how to grow parsnips.

One of my favourite memories of this era is pulling my Range Rover bonnet-to-bonnet with a VIP's Range Rover, lifting the bonnet and using my jump leads to resurrect his flat battery. He sat in the back seat with his jaw on his lap. I was dressed in a very small, lacy black cocktail dress and the (also Welsh) chauffeur stood back and let me enjoy my moment of glory with a knowing grin on his face.

My Range Rover was the old classic model, complete with a well-used tow bar and spattered with mud. It was my work vehicle and had my gardening tools in the back. The Range Rover we gave the kiss of life to had been the latest model, black, sleek, shiny and, which I noted quietly, was also a works vehicle with 'tools' in the back!

I had more regard and respect for my Range Rover and it's abilities than I did for my husband. It was more reliable. The marriage hadn't all been bad. There had been good times, just not enough of them to justify trading any more of my life for.

Marriage was obviously not my forte.

Work was definitely easier than relationships. I vowed to concentrate on work from now on. I would stick at what I knew.

Ironically, during this time I was also working with the boxer, Joe Calzaghe, on various campaigns for the wonderful charity, Beat Bullying. Bullying is not just confined to schools, nor is it confined to physical blows. It comes in many guises, to varying degrees, permeates many environments and is sadly overlooked and dismissed far too frequently.

I embraced the opportunity to talk and offer a little reassurance to schools and groups based on my own extensive experience of being bullied and discussing the part I had played in letting it happen.

Rather miraculously, I was slowly realising that the very episodes and behaviours I had struggled to understand and reconcile were now proving to be blessings that could be used to give strength and reassurance to others. I wasn't the only person going through these things. Far from it. The less something was discussed the more common it would probably be.

44 .

Bushcraft angels

'And those prayers were answered within 24 hours in the most beautiful way you could imagine.'

At the end of April 2014, just one month before my attempt to be the first person to run across Wales barefoot, I was handed a miracle.

Albert Einstein had claimed, "There are only two ways to live your life. One is as though nothing is a miracle. The other is as though everything is a miracle." I was always leaning toward the latter.

My 'Achilles Heel' of the challenge had been finding my way across Wales.

It was such a fundamental part of the challenge, however. I had trusted the universe and retained faith that a way, or more to the point, the way would be shown.

I had not had any further response from various people who had pledged their help in various guises. Everyone was busy; there wasn't enough time to factor in my soon-to-happen challenge. I was reminded frequently that this sort of project normally takes a year or even two, to organise. Normally.

Friends were also falling by the wayside. Many were unable to understand or condone my single-minded path and had got fed up with me refusing reunions and not replying to social calls. I regarded it as an organic pruning process.

The less I interacted with humans the more I turned to the ethereal world.

I asked. I prayed. I asked and prayed again. And again. With the volume turned right up. I asked with intense gratitude and sincerity and I asked again with lightness and with good humour. I asked in wonder and I asked with knowing. I kept asking. I kept praying.

"Those prayers and requests were answered within twenty-four hours in the most beautiful way imaginable. No - it was actually beyond imagination."

It went like this:

The Saturday morning of the May Bank Holiday weekend, I turned out in the wind and rain to join a local group supporting World Tai Chi and Qi Gong Day. I didn't want to go; I had a lot of things I felt I ought to be doing instead and to go and spend an hour or so celebrating universal energy didn't seem important, until I checked in with myself.

Of course celebrating universal energy was the priority.

I went. It was rewarding on all levels. The energy was tangible, exhilarating and as I left I silently gave much gratitude for the experience.

I went on to get animal food from the Mill where the staff immediately asked about my barefoot endeavours. Still buoyed up with universal energy, I openly admitted that it was the route that I was struggling with. It was a major issue and I was embarrassed that I still hadn't resolved it. Rose suggested that I popped into the local outdoor shop, Adventure Gear, to ask about GPS systems. I explained that I had thought about that but I knew nothing about the systems and was reluctant to add to my frustrations. Simply having an extra gadget, which I couldn't use and didn't trust, to carry wasn't going to help an awful lot.

"Go on," she said, despite my negativity. "Just find out. Do it now."

So I did.

Behind the counter was another lovely Rose (what are the chances of two 'Roses' being so instrumental in this tale?) who had been following my progress in the press and with whom I had even had a very brief conversation with while running barefoot through town one day. I had liked her.

As I trudged into the outdoor shop, Rose greeted me with, "Oh, we've just been talking about you; we have a new barefoot shoe in stock and obviously your name came up!"

Obviously.

She asked how it was all going and sighing heavily I admitted that, ridiculously, my main stumbling block was the route. I mumbled something about GPS and then rather randomly blurted out, "Ideally what I need is someone to walk the route ahead of my run and mark it all out for me to follow. Like Hansel and Gretel with breadcrumbs. But little laminated signs instead of breadcrumbs."

Rose stared. I assumed she was thinking how stupid I was to be attempting to run across a country in less than four weeks and have absolutely no idea or inclination of a route. It was madness.

Instead she replied, "I think we could do that."

She explained that she had actually signed up to do the Rotary Walk Across Wales the previous year but had overslept, missing the bus that took the walkers to the start. She added, "I know the area well too; I used to live in Machynlleth. I keep meaning to do it but just haven't got around to it."

And that's not actually the pure, golden, magical bit. This is:

Rose went on to explain that she and her partner Gary were getting married on the Tuesday before my run and had been looking for a meaningful Welsh walk to undertake as their honeymoon. "So", she smiled, "it would be great to walk across Wales and we can put the markers in place for you to follow."

When these magical synchronistic things happen I get a bit light-headed and am apt to ramble. I started to waffle about having maps but not even having enough information for them to follow when she added serenely, "Gary is an ex-Marine. He does all that orienteering stuff. He loves it. He'll sort all that out."

It also turned out that they were starting up a bushcraft company and were both going to be at a local 'Wellbeing Matters' event which I had been asked to open the following morning.

They would both see me there. Of course they would, the universe had it covered.

I met Gary, who was just as cool as Rose, and who simply added, "Yeah, it'll be great. Works for us all! I love that universe stuff. It'll all work out. Rose is manifesting at a rate of knots at the moment, we'll leave it up to her."

We all spoke the same language.

And that is why I talk to spirits.

45.
Paws for thought

'She took her last breath in my arms.'

Throughout all the peaks and troughs of my mid-life I had one constant companion. Reliable, non-judgmental, loving and loyal, she entered my life as I entered my thirties. She was my dog. My little Westie, Tippy. She had stayed with me after my first divorce, my four-legged angel. She was also the first dog in South Wales to have a Pet Passport.

I had a tank bag made for my motorbike and she travelled with me on numerous motorbike excursions. She travelled down through Europe with me, came to work with me and made numerous house moves with me. She even came to the top of a mountain every New Year's Eve to welcome the New Year in. We would recce our route a couple of days beforehand, then cwtch up together on top of a mountain looking down on the fireworks and wishing wonderful things for the coming year. I knew that if I could muster the courage and motivation to do that each year then I could muster the same determination to deal with what lay ahead during the following year.

When it wasn't feasible for her to be by my side, she would wait patiently in the truck for me.

The only thing she didn't do was appear on TV with me. The producer had advised against it.

"It's a crazy world," she had said. "If you have something that you value as much as you value your dog, they become a target. You may find some nut kidnaps her or just threatens to."

She was kept away from the TV cameras but loved her time in front of the other cameras and featured on my first calendar and in most of my media shots.

We were inseparable.

In the summer of 2009 she had developed a cough and the dreaded relationship with the vet began. She got more and more out of breath and tired easily. She would also have the most terrifying little black outs occasionally. Determined to honour our partnership, I experimented with several options of carrying her. Taking the rubbish to the end of the lane was easy; she would sit happily in the wheel barrow. She also loved her Mayan sling, designed to carry small children but it played havoc with my shoulder.

It was OK for short walks and participating in a charity walk in Cardiff with David Cameron. Despite her ailing health she was still determined to protect mine. When Cameron strode up to us, no doubt to make the most of a patting-dog-photo-opportunity, and commenting on the cuteness of me carrying her, she snapped at his outstretched hand, missing his fingers by a whisker.

She would have made TV for all the wrong reasons.

I eventually converted a back pack for carrying children for her. With a little cushion and her blanket she was as comfy as anything and proved an added bonus to my fitness training as we continued to hike over the Welsh hills, with her enjoying the views over my shoulder.

At the end of the summer she was diagnosed with Westie's lung. The vet explained it was incurable. So how long did we have? Twelve months, surely?

"A month," was the stabbing reply.

A month.

Four weeks.

Thirty days.

I was devastated.

She was just over fourteen years old and we had been inseparable for fourteen years.

I was in the middle of a second uncomfortable divorce and Mum was back in hospital; I was exhausted. I couldn't bear the thought of losing my best friend.

I virtually gave up work and I'm sure any work I undertook was hardly worth doing. People were incredibly understanding. I spent every day with Tippy. Every moment. We camped in the dunes of our favourite beach, Barafundle Bay. We climbed our favourite mountains; I climbed, she took in the fresh air and views over my shoulder, becoming breathless even in the back pack.

She was fading and so was I.

I cherished every day we had and yet dreaded it too, as it was one day closer to the inevitable. I prayed I wouldn't have to take the final decision. I prayed for a miracle. I prayed for help. I prayed for strength. I just prayed.

On the morning of the 22nd October 2009, she stumbled out into the garden, disorientated and distracted. She pushed under a beautiful wild fuchsia bush whose delicate white blossoms bowed silently like little angels. I followed her, sat on the soil, scooped her up onto my lap and cwtched her close under the little fuchsia angels.

She took her last breath in my arms.

My tears fell amid the autumn leaves for a long while.

Distraught, grief stricken and almost emotionally defeated, I hurt like I had never hurt before.

And that was the darkest, coldest, hardest, most heart-breaking moment of my whole life.

46.
Two loved feet

'My toes were shredding.'

With my challenge weekend looming, my bare feet were receiving much attention. They also received an invitation to visit chiropodist, John Cox, in Brecon, who was rather bemused to hear about my challenge and intrigued to find out what condition my feet were in.

He had heard about my barefoot journey on Radio Wales when I had shared my rather unconventional methods of acclimatizing my soles within the limited time I had. I had for a while been wearing my work boots without socks, which had taken a little adjusting to. The next stage had been to add a little grit to my boots in the morning. Where most people were shaking grit out of their boots I was putting it in. It had intrigued and amused the listeners, the host and John.

I sat tentatively in his chair and awaited the verdict. He deduced that my toes were having the worst deal. The bad news was that they were peeling, fraying and rather bruised; shredding was the technical term. My toes were shredding. However, the good news was that in general, my feet appeared to be quite happy and coping admirably.

John recommended little gel filled digital caps to protect my toes should they find it all bit much during the actual challenge itself. He advised, "they may mean you are able get a few more miles out of your toes."

He was actually more intrigued by my Sockwas, the latest barefoot shoes I was testing courtesy of fellow bare footer, Tracy Davenport. Tracy had also been a great help with nutritional advice and is a natural motivator. As she had predicted, I loved my Sockwas.

They were an absolute dream, making those hazardous terrains manageable. They were, as the name suggests, little more than socks and while you can feel every little stone and stick through them, they protect against puncture wounds. They made my Vibrams feel like wellingtons. If you are keen to try bare footing and want to take it gradually, then Sockwas are a good way to get started.

I had also tried the infamous Lunas, designed by Barefoot Ted, inspired by the Tarahumara Huaraches, which you can read about in 'Born to Run'. I was skeptical as to how practical or comfortable it would be to run in a pair of sandals but, as thousands of other runners have already discovered, it is exhilarating but I still couldn't quite get on with them. A little bizarrely maybe, I actually felt more self-conscious running in sandals than with bare feet. They also have quite a thick sole compared to the Sockwas so I decided to keep them for every day, conventional use when footwear was more appropriate than bare feet.

Around this stage, I also noticed that my feet had actually grown or spread. They were now a good half size, nearly a full size, bigger than when I had started training. They had changed shape too and were a lot wider across my toes. It is a foot's natural shape - think of a baby's foot, they are far more triangular than an adult's. It's what happens when feet are not entrapped and curtailed in shoes. I loved my new feet.

Ironically, it wasn't until I had completed my challenge that I was introduced to my very favourite barefoot shoe of all. Made from chainmail, they are Paleo Paws.

To call them barefoot shoes really doesn't do them justice. They are so very much more. They offer the luxury of being able to feel the earth beneath your feet and also to have the reassurance of protection from puncture wounds. And they are quite simply beautiful.

I was immediately impressed as I opened the metal box they arrived in. The presentation is exquisite and demands your full attention. You immediately know you have something special in your hands. They feel unnervingly familiar; they resonate.

Creator, Jorg Peitzter, explained he had created them to enable bare footers to conquer the more unfriendly terrains they may encounter whilst retaining that all-important connection with the ground.

My passion for the Paleos appeared to be of a primeval nature. They took me to a primal place; they nudged a latent memory of a primal state. They made me feel warrior-like, a little invincible even. And that is the perfect antidote to the vulnerability that being barefoot can evoke. They are quite simply chain mail magic and a reassuring option to being bare foot in less favourable environments.

I actually believe that many of the experiences and benefits of spending time barefoot are enhanced by the Paleos.

I adore them.

47.
Miss Understood and Mr. Nice

'... fuelled idle (and some not-so-idle) gossip that I was 'losing the plot'.'

The autumn of 2009 was bleak. I had just lost my dog and my husband. I cannot begin to explain how much I missed my dog.

Losing her broke my heart. And through the fractured fissures flowed all the agony, hurt, angst, hatred, frustration, pain and fear which I had carried all my life.

I didn't know where to turn.

I turned to the spirits.

Before her passing, I had spoken regularly with Tippy, especially about the painful situation that loomed. Thankfully, I found our communication continued uninterrupted and effortlessly after her death. We had the most profound conversations.

Of course I didn't tell anyone. I was distraught and exhausted. My divorce was taking far longer than was necessary. The more I retreated the more ferociously he attacked. I readily agreed to take any blame for the breakdown of the marriage in order to speed the process up. I had lost my impetus for work, my familiar hiding place. I felt that I was hanging on by my finger tips and I was losing my grip.

It transpired that what I thought was a breakdown was actually a breakthrough.

I wrote all of the conversations I shared with Tippy in a journal. I worked closely with a spiritual friend. She helped me fine-tune various innate abilities. I learned things that I never thought possible. I had always believed there was more to life than we realised, than we talked about, than we embraced. Now I knew it to be true.

And with the enormous comfort it brought, also came an inevitable increase in solitude. I became less and less able to engage in small talk, to endure lies, gossip and nonsensical bullshit. I felt more and more alienated from the mainstream and yet more and more connected to what mattered - what really mattered. Not only did I no longer want to continue the charade of the incongruent social scenes, I couldn't. I just couldn't do it.

I turned down social and professional engagements, public speaking and appearances and became more and more reclusive. I pursued my interests in healing and spiritual matters. I trained in lesser-known practices like Flow (energy work) and Zoopharmacognosy (how animals self-medicate in the wild), dowsing and working with medicine wheels and labyrinths as well as learning more and more about herbal remedies and folklore. I loved it. It felt so right. It felt like I had returned home. I had been away too long.

I approached Cardiff Prison with a view to introducing labyrinths into the exercise yards for the inmates to benefit from. It was the first time I had been there since visiting with Mum 30 years ago.

A lot had happened.

After several meetings, it became apparent that labyrinths were not going to make their debut into exercise yards at that time although I did succeed in introducing outdoor exercise equipment. It was also an unexpected completion of yet another cycle.

Ironically my rejection of public commitments and increased interest and commitment to the more spiritual aspects of life, fuelled idle (and some not-so-idle) gossip that I was losing the plot.

I was of course finding it, finding myself. It seemed that not everyone was as happy as I was about it. Personal growth helpfully has its very own pruning process.

The last media appearance I made was on the Paul O'Grady Show. Dermot O'Leary was the guest presenter as Paul was ill. I was asked to appear as a guest expert as their regular gardening expert, Christine Walkden, was also ill. I was to show Dermot how to plant up some unconventional containers with veggies, in an attempt to encourage him and the viewers to 'grow their own'.

I wasn't sure about it. I didn't feel up to it. I decided to do it and I would make it the last. I would also be authentic. Be myself. No more acting. No more front.

I was treated like royalty. I wore a simple vest top, my beloved jeans and my trusty work boots. That was my natural authentic style.

I enjoyed lunch with fellow guest, American actor and professional wrestler, Dwayne Johnson, aka The Rock. I met Whoopi Goldberg and Paul Merton in the Green Room. I enjoyed banter with the crew and the catering staff before the live show and then sat quietly with The Rock behind the set waiting for my cue to go on. It was being recorded live.

Dermot suddenly bounced in by my side, gave me a big kiss on the lips and said, "I've been hearing how wonderful you are. Lovely to meet you."

The Rock offered a high five. "Amen to that. You rock, girl."

I floated out in front of the cameras; instead of being introduced as a gardener from Wales, as had been agreed, Dermot introduced me as Wales's Sexiest Gardener and affirmed, with a wink, that he could see why.

The show passed by in a mist of magic, laughter, fun and flirtation. And as I sat on the train coming home I gave my gratitude to the Gods and reaffirmed that it would be my last public engagement.

It was the best possible place to end. For the moment. Little did I know I would return to that arena in the best possible circumstances.

I continued to interview celebrities for my Green Scene column in the Western Mail but politely declined any more TV or radio work and any invitations I received directly from the celebrities themselves - except the ones which were too tempting. I had a lovely lunch with Robert Plant, enjoyed banter back stage with Francis Rossi and Rick Parfitt and met with Piers Morgan, Roger Daltry, Ringo Starr and Ricky Gervais at the Chelsea Flower Show.

And the universe delivered the odd unexpected event too. In my late teens, I had looked after a dog and a house for a friend of a client and who lived next door to Pinewood Studios. I had spent my time there rehabilitating a rescue dog she had just re-homed. She had been extremely grateful. Out of the blue, she invited me to supper for a catch-up via our mutual friend. I was in London anyway, so accepted expecting a casual exchange of news. Instead I spent the evening sitting next to another friend of hers, Hugh Grant.

However, the contrived social obstacle courses - more commonly known as parties and dinners - held no appeal whatsoever any more. I simply wasn't strong enough. I still felt incredibly vulnerable and fragile. I was walking the path of authenticity and it was, in parts, painful. However, I still loved chatting to the celebrities on the phone. I was given a completely free hand to write what I wanted as long as there was a Welsh connection.

I would respectfully enquire about their new film/record/show/book before admitting that my column was actually called the Green Scene and therefore I wondered what their own experience of gardening or the great outdoors was. By this time, I had gained their trust and confidence and they relaxed as they discussed long-forgotten childhood memories or retirement plans which involved getting back to nature. It was lovely. They were lovely.

When my email account was hacked, sending a fraudulent plea for money to all those in my address book, Bill Bailey was one of the first to call to warn me. Julian Clary mulled over having two of my little pigs and Rhod Gilbert helped me with numerous Welsh-based fund raising stunts. I answered my phone to the dulcet tones of Ben Fogle, the chirpy tones of Carole Smiley and the chortles of Christopher Biggins, all returning my requests for an interview. They would comment on the chickens clucking contentedly in the background or patiently wait as I turned the oven down, postponing my supper to chat to them. My authenticity seemed to put them at their ease.

I became friendly with many of those I interviewed. They would email for advice on an aspect of gardening or how they could maybe keep bees. Or just to say hello.

But possibly one of the most unexpected friendships was struck up with Wales's most Infamous celebrity, Howard Marks.

I first heard about Howard Marks while in my late teens; a lot of my friends had enjoyed the same recreational habits as Howard and both he and his antics were respected and revered by them. Despite my own dislike and disinterest in drugs, there was something about the way they described him that struck a chord. I decided that I would like to meet him. And that was in 1984.

Ten years later and moving in a totally different circle of friends, a high-flying, financially-successful admirer gifted me Howard's autobiography, 'Mr. Nice.' He correctly predicted, "You'll find this guy interesting."

One of the few compliments my Dad ever handed to me was that I was good with people - when I wanted to be. He told me that I would always be able to mix with princes and paupers and all those in between.

Maybe that too had been more of a prediction than an opinion.

I have always been happy to run with both the hares and the hounds.

A lot of the people I most admired and respected as a young adult had rather unconventional and often controversial lifestyles. I never felt pressured to partake of any of their particular indulgences but, nonetheless, I enjoyed their company. I had also earned the reputation of having a sixth sense. If I said it was time to go, it was time to go. More than once, I would be heading one way along a narrow country lane as police sirens announced their approach from the other.

While I didn't condone the particular past times of these acquaintances, I did enjoy their attitudes, wisdom and company.

I met Howard a quarter of a century after deciding I wanted to. It was worth the wait.

There was an instant and mutual intrigue and respect when we met. He is one of the most intelligent and articulate men I have ever met.

I don't like drugs. Of any type. Prescribed or otherwise.

"I don't see why you need any mind altering substances," I would preach. "Why can't you just be yourself, without all the smoke screens?"

Yes, I was asking Howard Marks, often described in the press as, 'the most sophisticated drugs baron of all time', why he had to take drugs; why didn't he just give them up and be himself?

Howard would calmly and charmingly point out that my addiction to work and exercise was inducing, and taking advantage of, several mind altering substances, adrenalin and cortisol.

"Yes but that's natural," I would argue.

"Oh and cannabis isn't?" He would raise a disheveled eyebrow.

We discussed and debated way into the small hours, him consuming copious amounts of red wine and me drinking jasmine tea.

"Nice high?" he would ask with a little slice of sarcasm, as I bounced in from an early morning run.

He would explain tirelessly how taking drugs heightened every experience for him, from listening to music, to writing, and he promised that with my level of intelligence and creativity I would get the same enhanced focus and enjoyment if I were to partake.

Keen to emphasise my lack of interest in drugs, I struck up a deal with him.

I would try drugs when he had run a full marathon.

That way we were both comfortably wrapped in the knowledge that neither of us would have to step out of our chosen zone of comfort.

I adored his philosophies. As I struggled to analyse a particular behavioural pattern, he would interrupt, "Does it matter?"

"Yes of course it does," I would reply and continue to attempt to justify my defence.

"Yes but does that really matter?"

"Yes - because..."

We would continue with this conversational tennis until slowly I realised and accepted that actually, when you peeled the layers back from any challenge or problem, 'No it didn't really matter'. Not in the great scheme of things.

He shared his fear of spiders and of death.

"Does it really matter?" I asked.

"Of course not" he grinned, "that's why I don't usually talk about it."

I feel incredibly privileged to call him a friend. I like the person I am when I am around him. I step up, paying close attention to my behavior, motives and grammar.

I once came back from an early morning run and explained, "I don't know what it is but when I spend time with you, it's like taking a net curtain down from a window or cleaning a dirty windscreen. My clarity and focus just improve. Everything seems clearer and brighter somehow. I love it. Thank you."

He smiled. "It's reciprocated."

I quickly adopted some of his philosophies, his appreciation of pedantry, his promotion of disengaging with unsatisfactory situations and dialogues, and also his love of the Groucho Club in London, which he introduced me to. I adore the place. Steeped in years of hopes, dreams and hazy fantasies, it evokes the most enlivening energy. I regularly muse over the most magnificent memories of time spent there and contrary to the famous Groucho Marx quote, would dearly love to be a member of the private members club.

I also adopted his lack of concern as to whether he was accepted or not, understood or not and embraced or not.

I quietly accepted and started to trust the friendship of a small group of girlfriends; they include a Shaman, a healer, a medium, an author and several other kind hearted souls with an interest in and respect for all things ethereal. We were affectionately referred to as witches and took it as a compliment. We were an eclectic bunch.

Occasionally we would meet up in a conventional setting to catch up on various adventures. We met to celebrate the Winter Solstice in 2010, a time when everyone else was celebrating Christmas. Amid the chatter, arose a desire for me to explore pursuing another challenge. A meaningful one; one to celebrate and honour my new-found authenticity.

How about a barefoot 10K run?

And that was how it all began.

48.
Flower power

'Stephen Fry was charming;
as was Benedict Cumberbatch'

I am passionate about motivation and inspiration, the importance of them in life and the numerous sources of them. Why do some people seem to be able to access them easier than others?

I read that Robin Williams claimed that, "Courage is one step ahead of fear."

I read a lot.

While training for my challenge, I devoured many books, hungry for inspiration and if I'm honest, maybe even a few short cuts. I was searching for the secret. That secret thing that you think other people have discovered and have access to. They, of course, are all thinking the same of others... and of you.

I read the fabulous 'Finding Ultra' by Rich Roll. Pure inspiration in paperback. I read 'Wild' by Cheryl Strayed, 'The Ragged Edge of Silence' by John Francis. I re-read Malcolm Gladwell's 'Outliers.' I hadn't reached my 10,000 hours of barefoot running but I would. One day. I swotted up again on Florence Scovel Shinn's 'The Game of Life.'

I read and I ran. And read and ran.

I got back in touch with Rosie Swale Pope. I had interviewed her for my column a few years ago. She and her many achievements are simply jaw-droppingly amazing. She was wonderfully supportive and even said she would see if she could be at Ynyslas to watch me complete my challenge. She is a beautiful, humble lady, with little need for recognition and a huge heart.

And the other Rose in my life was proving to be equally as awe-inspiring. Rose and her partner, Gary, the couple who had offered to spend their honeymoon marking out my route across Wales, had also adopted the role of being my official back-up crew. They had helped me clarify the few outstanding decisions. I would run over two days, starting on the Saturday morning and finishing at the coast on the Sunday afternoon. God Willing. And I would run alone. Rose and Gary would leave Ynyslas on the Wednesday before my run, walking and marking the route in reverse, as their honeymoon. They would then meet me at Anchor and drive my truck as a backup vehicle with supplies as I attempted my challenge. They would follow me where vehicular access allowed and wait for me at designated locations when it didn't.

They are both believers in universal energy and spirit guides as well as being incredibly pragmatic and practical. It was an absolute delight to collaborate with them for the challenge.

It is all that being barefoot encompasses - trusting your divine path! I am still very much that loner but I can still interact with like-minded people, my tribe.

I had been gifted with the most amazing, gentle little team from the universe and in gentleness lies great strength.

By the beginning of May, my challenge month, I had shaken down in to my preparations. Less and less time was spent worrying about what still needed to be sorted out and more time was spent in the moment. It worked. That's what happens when you relinquish control and surrender.

It was also that magical time of year when I get to go to Chelsea Flower Show and record interviews for BBC Radio Wales. It is the professional pinnacle of my year. I love it.

Mulling over possible interview opportunities, I mused that I would like to meet Stephen Fry; I had read his revelations about bi-polar with interest. There was no telling who would be at the Flower Show. He hadn't been there in previous years.

He was there this year.

And he gave me a wonderful interview. I caught his eye amid a large media scrum and he picked me out to talk to, much to the frustration of much more qualified and hardened reporters. He was charming. We had our photo taken and I whispered my gratitude to the universe.

I also met Benedict Cumberbatch. As I came out of the ladies, he came out of the adjacent gents and complimented me on my flowery boots, remarking that his mother would love them. As we came back onto the show ground, he was immediately surrounded by the familiar media crush of cameras and microphones.

I had had no idea who he was and had missed an interview opportunity. I relayed the oversight to my brother. He grinned and said, "No shit, Sherlock!"

In addition to my annual social highlight, I was also enjoying higher levels of energy and fitness and regular rewarding runs. As the sun rose one morning, I ran six or seven miles over the moors to The Chartist's Cave, fully intending to then turn around and run back. My spirit, or maybe the other spirits, had other ideas and, like Forrest Gump, I kept on running for another six or seven miles before eventually turning around and running home. Yogi was with me and kept pace every step of the way. She was becoming an ultra-running dog, bless her heart. Dogs are miracles with paws, bare paws.

As I ran I mused that I hoped that my journey will inspire others not only to spend more time barefoot but also to push their own boundaries and step out of their comfort zone into magic meadows.

I had seen a lovely quote written on the back of a camper van which read, "Life begins right at the edge of your comfort zone." How true.

With challenge day just around the corner, I reflected on how it had been the journey rather than the challenge itself that had enabled me to make such important and congruent changes in my life.

How I had come to accept that it was OK to prefer solitude and not to want to be involved in that which didn't make me happy or serve me in some way.

At last I felt comfortable in my own skin. I didn't feel a freak or alienated any more. I felt proud to stand alone and to be different. I didn't need the misleading safety of a crowd to hide behind. I no longer felt the need to please others. I was happy and comfortable pleasing myself. If I was happy then those who genuinely care about me were happy too.

I wrote: "I am who I am. I no longer feel that I am broken. That I need to be fixed. No longer searching for the secret. I am all I need to be."

Of course I could and would choose to continue to grow and develop and learn but it was no longer essential in order for me to become whole. I was already whole, already complete.

Let go of your fear and risk finding happiness. I believe we are taught to attain and retain. We are taught to work hard to get those qualifications, that job, that spouse, that house and then work even harder to keep it all.

If we were all brave enough to let go, then we would free things up for those who would find joy in them. And we would find our own joy as someone released it for us. Stop hanging on.

The secret is to let go. Let go of your fears, of your expectations, of ideals and beliefs that no longer serve you. Detach from the outcome. Go with the flow, trust and be trustworthy.

While training for my challenge, I had revisited a process called PSYCH-K which was created by Rob Williams and endorsed enthusiastically by Bruce Lipton. Again, as if by magic, the only facilitator in the UK lives in Cardiff and offers PSYCH-K through her company, Yearning for Learning. I highly recommend it and her.

It is a powerful tool to help you change your debilitating belief patterns into beneficial ones. You don't have to spend seven months running barefoot training to run across a country to evoke these changes, the PSYCH-K courses are a shorter and easier method of exploring similar benefits.

It only takes a second to change your mind, a few seconds to change your belief and those shifts will change your entire life. Choose to be happy. Choose to follow your dream. Choose it now.

I was just a 47 year old gardener who had decided to see if she could get her mind body and 'Spirits' to work together, to collaborate, to achieve something meaningful.

There had been emotional 'casualties' and I had made several sacrifices. I had learned that I didn't have enough time to spend with people that I liked, so I certainly didn't have time to spend with people I didn't.

Nigel from Blue Mountain had reassured me, "Challenging oneself is a lonely old game. Not everyone understands it and even fewer want to buy into it."

Training had included running up mountains and reaching places - geographically, physically and emotionally - that even I only thought about a few months ago.

Now I was accepting the invitations softly whispered by those beautiful Welsh hills and rather bizarrely much preferred to run up those challenging hills than along the flat banks of the canal or gentle undulations of the open moorland. Yogi, my second little Westie who had stepped into Tippy's paw prints, had become my off-road running buddy. I was so often inspired and encouraged to run a little further by that little dog with the largest of hearts and the stamina of a lion.

It was a beautiful life.

49.
The big weekend

1st JUNE 2014
I wrote in my diary:

I did it!
I bloody did it.
I have just run into the history books.
I am officially the very first person to run across Wales barefoot.

Whilst writing my blogs, I had often idly wondered what it would be like writing the one after the event. It was harder than I thought. I didn't know where to begin.

I have so many people to thank but I am aware you would probably rather read about the actual pain, pleasure, more pain and practical aspects of the run itself. Therefore I will write a special gratitude page later listing the many fabulous collaborators and supporters and now will try to give you a run-down (pun intended as always) of the Challenge itself.

I met Rose and Gary in Anchor on Friday evening as they completed their walk across Wales, marking the route for me to follow as I ran over the weekend. I asked Gary how it had been and he replied, "how much do you really want to know?"

Now that's the type of answer I like. I had an option but of course I also had to know the truth.

"OK," he shrugged, "it's going to test you to your limits. There are some very difficult sections."

Rose added quickly, "There are some nice bits - and great views."

And that is just one of the reasons this newlywed couple were heaven-sent - they were the absolute best support team I could have wished for.

I remember feeling quite numb on Saturday morning, emotionally not physically. I had managed to get myself into a good frame of mind; not too anxious, not too excited and as downright determined as I had ever been in my life. I had worked hard, hopefully hard enough and so had a lot of other people. Now I just had to do it.

I set off from Anchor at 9am straight up a tarmac hill and at a light dog-trot. I find the first couple of miles of any run hard, just shaking down into it - adjusting clothing, pace, stride and just generally sorting my thoughts out. The quicker I could get into my stride today, the better. After a quick check with Rose and Gary at the top of the hill I headed off cross country for my first solo section.

Gary and Rose had decided to follow me in my truck for the road sections to keep me protected from traffic and then would meet me at the end of the cross-country sections they couldn't access with the truck.

We had an intrepid photographer, Mike Erskine of We Are Here on board too, following Rose and Gary in his car. Mike turned out to be another godsend. As well as great company, he would walk back towards me along the off road sections to get the photos he wanted. As soon as I saw Mike and his camera, I knew I was nearing the end of a cross-country section. Incredibly self-sufficient and innovative, he chose some wonderfully elevated and camouflaged places to sit and wait and quite often I ran past him without seeing him. I teased him that he would make a great sniper if he got bored with using a camera to shoot with.

Back en-route and muddy farm tracks morphed into stony paths which in turn led to grassy fields and boggy marshland with long stretches of pitted tarmac in between. If variety was the spice of life, this was a madras.

Gary and Rose would meet and greet me at various points, with Mike - his face constantly covered by his camera - and slowly, minute by minute, mile by mile, I edged ever closer to the coast.

The first six miles were enjoyable; I was running high on adrenaline and anticipation. The next six were OK. Just OK. And then I had a pain in the top of my right foot. I was just twelve miles in to a fifty-two mile run. I employed all the psychological tricks I knew and kept running; that pain was to keep me company for the rest of the run.

It was at the wind farm section that I first got lost; there was no vehicular access so Gary and Rose had driven the long way around and planned to meet me as I exited the moorland at an estimated and agreed time. I missed one of the marker signs and ended up right at the end of a dusty, flinty track in a stony cul-de-sac. There was nowhere to go, only deserted moorland. I hadn't seen a marker for miles but assumed, at low ebb, that they had been removed. I was tired and also a little paranoid it seemed. After jogging back along the track to the first intersection hoping to pick up a sign (from God or Gary), the enormity of the situation dawned on me. I was in the middle of nowhere, with no mobile phone signal and nobody to turn to.

After a long half hour of trying to find an Orange-given signal and a God-given sign, a man on a horse galloped over the far horizon. Thankfully he was heading my way. God had responded if Orange hadn't.

He called out against the wind, "Are you Lynne? I've come to rescue you."

You can't make that stuff up.

All my life I have wanted a knight in shining armour to gallop up on his horse and rescue me and here he was. He had left his armour at home in favour of a tweed jacket but as for the rest of the manifestation, it goes to show that you get what you wish for. Eventually.

It transpired that The Team had met Michael Brennan out on his hack and explained the predicament as I had over-ran my estimated time of arrival. Michael had offered to ride on to find me as he knew the area; it had taken him an hour of searching but he had found me. I was both grateful and relieved as I jogged out of the wind farm alongside his horse, Joc, and his dog, Pip.

Incidentally, in Welsh Brenin means 'king'. Knight? King? It was all the same at that magical moment.

That rather surreal episode kept my mind nicely occupied for the next couple of hilly miles and to be honest there are awfully long stretches of the run that I can't recall very clearly at all. Any running style I had started with had been long forgotten. No longer was I concentrating on lifting my knees and relaxing my shoulders, now I was just getting one foot in front of the other... and repeating.

I was still running at 7pm and didn't think the day would ever end. It did. At about 8.30pm and after another monumental hill climb.

For the last couple of miles Rose had encouraged me beautifully by shouting, "Just another mile or so." I had run thirty miles which is what we had aimed for; it felt more like sixty and I honestly couldn't have run another mile that night. I was right at the end of my capacity. My right foot was still playing me up, the soles of my feet were burning and my legs felt like lead.

And I had another twenty-one miles to do tomorrow.

I gave gratitude to the spirits and to the countryside that I was running through and collapsed in the back of the truck.

As Ralph Waldo Emerson recommended, 'Finish each day and be done with it. You have done what you could. Some blunders and absurdities no doubt crept in; forget them as soon as you can. Tomorrow is a new day; begin it well and serenely and with too high a spirit to be encumbered with your old nonsense.'

I was spent for that day but the universe was still delivering; we stayed in the most beautiful B&B in Dylife with a fabulous homemade dinner and the most wonderful bath and bed. And we had been blessed by an equally welcoming family in Felindre on Friday night. I could write a blog on each of these perfect places to stay, each with their own individual attributes. They were just the best.

Medina, from Brandy House Farm in Felindre, was particularly interested in my barefoot journey as she kept her horses 'barefoot', or unshod, for trail riding and explained that the horses encountered similar transitioning issues from shoes to bare-hooves. I shook my head at the serendipity of it all.

And at Bron Y Llys, where Maya is quite possibly the best cook in mid-Wales, we watched hares play outside, under a starlit sky, whilst we ate. It had been the first dry night for over three weeks. So much magic on so many levels.

We were back on the road, or stony track to be precise, by 7 am on Sunday morning, conscious that I would be slower today and we wanted to arrive at the end at a reasonably civilised time.

Setting off that morning was tough. My legs whined, my feet moaned and the rest of me wasn't an awful lot more cheerful. Yet within a mile I was running along the edge of a beautiful forest under magical mists which were predicted to lift by mid-morning. I had seen a fox and several more hares and was accompanied by two red kites overhead.

I felt completely enveloped in nature and carried by the elementals and nature spirits.

It was a section I will never forget. And that magical energising experience was to be even more valuable as the rest of the route became more and more gritty and grueling. The pain in my foot was almost unbearable at times. I had to limp downhill to try to alleviate some of the pressure. That put a strain on my left side. I prayed and not for the first time.

Noting the ebb of energy and enthusiasm, Gary accompanied me though a ridiculously hard, steep and unforgiving forestry section. I remember only seeing his back the whole time but he got me through it.

After the long stretch of tarmac that followed the forestry, Rose insisted on buddying up with me to do the last monumental stretch of cross country, the coastal path. I didn't refuse. If I had, I may well be writing a different blog; probably blissfully unaware of her input, it was Rose who got me through that crippling part of the route. Every single step hurt; it was largely down-hill which was a nightmare for my ailing right foot. The path was cruelly uneven and stony - every single stone felt like a golf ball under my foot. It was horrible and it felt personal.

Rose remained upbeat, pointing out the coast. At last the finish was in sight. Still fifteen miles away, but in sight.

Surely I couldn't fail this epic challenge with the finish and coast in sight, could I?

No, I bloody well couldn't. After a quick 'Scooby Snack' and attitude adjustment at the end of the track from hell, I had just two miles of tarmac to go conquer before running to the finish on the beach at Ynyslas. As I munched my apricots and pumpkin seeds, a cheerful family approached and offered their congratulations and support. Lesley Jeffrys and her family had seen the markers on their walk and were intrigued to know more. They were kind and complimentary about the challenge and my efforts. That encounter was the just the encouragement I needed to cover the last two miles.

And what a hostile, unyielding and painful two miles of broken tarmac it proved to be. I tried to run on the faded but smooth white line along the edge of the road but it wasn't feasible. It was also covered with hedge trimmings and broken glass. I was forced to run the cracked tarmac with its skin shredding tendencies. With every short, painful step and not even daring to look up at the horizon, I chanted, "just one more step, just one more step."

Rose kept up her brilliant but inaccurate account of the distance remaining.

Turning off the tarmac and onto a short and unexpected stretch of stony track nearly reduced me to tears. I had to stay composed, focused and determined. Just half a mile of sand to go.

Gary and Rose drove on to the finish.

I was alone. All alone on the very final stretch of sand that would turn a corner and lead to the home straight and the completion of a very long journey.

That final section of sand was a gift from the universe; to be able to run into the finish and the waiting crowd with some degree of comfort and dignity was a sheer delight. My feet and my heart sang.

When I pull that last 500 yards up as a memory, it is as though I am watching someone else. The relief, the joy, the pride, the whole range of emotions blurs the image as a soft-focus lens would.

Amid cheers and hugs, I stopped running.

I had finished.

It was over.

I had done it.

I had become the very first person to run across Wales barefoot.

BAREFOOT AND BEYOND

And the journey continues. Which of course is a good thing.

The sequel, Barefoot and Beyond is 'under construction'.

My barefoot challenge may have ended successfully on a Welsh beach but I still found myself looking over the sea to the horizon. It wasn't the end but the beginning.

Of the next chapter – or, as it transpires, the next book.

I have been humbled by the amount of interest that has been shown by people wanting to know more about spending time barefoot and the benefits it can provide.

Journalists have been intrigued enough to share details of my unconventional path and some have also been inspired to pursue their own bare footing endeavours long after our time together.

There are numerous projects and possibilities (or probabilities as I like to call them) in the proverbial pipeline and life is good.

Of course life continues to be unpredictable and the path occasionally a little tricky to negotiate sans sat nav. Other people talk of learning curves; I seem to get learning hair-pin bends. But as the saying goes, "a bend in the road isn't the end of the road unless you fail to make the turn."

As yet I am still making the turns. And writing about them.

The road continues to be rocky in places but being barefoot provides more support and protection - emotionally. I love that paradox.

There are encounters with God, or a 'higher power'; you can choose your own title, but the experiences remain the same.

I wonder how many people will now be rolling their eyes and closing the book. I might have a few years ago. Now I look forward to sharing that particular encounter and episode in 'Barefoot and Beyond'.

Oh and the moon. I have been working closely with the energy and phases of the moon. If you doubt the ability of this incredible force just think of the effect it has on the tides and then remember that we are also 70% water. How could we not be affected by its waxing and waning? I have found it to be a strong guide and working with it to be extremely productive.

Of course the irony is that the joy of finally accepting my love of solitude and my own space has been ever so slightly compromised as I continue to encourage others to explore the concept for themselves.

I have been referred to as The Reluctant Guru, the Hermitess and even been called Feral. All of which I perceive as compliments. It's all about perception.

I am learning to balance sharing my time with others as well as retaining my own boundaries and quiet time which of course is essential for the development of myself and my journey, barefoot or otherwise.

Barefoot and Beyond offers more information on how to spend time barefoot, how to incorporate it into your life, using it as a therapy, a tool and a pathway to improved health and self-awareness.

Of course there are celebrities.

There are gains and pains.

There are challenges and achievements.

There are details of my next all-consuming challenge, already in the planning stage.

There is even romance.

And there is an extremely painful and raw loss.

I do hope you can join me on the next leg of the journey.

Lynne x